Anthracite's Demise and the Post-Coal Economy of Northeastern Pennsylvania

Anthracite's Demise and the Post-Coal Economy of Northeastern Pennsylvania

Thomas J. Keil and Jacqueline M. Keil

LEHIGH UNIVERSITY PRESS
Bethlehem

Published by Lehigh University Press
Copublished by The Rowman & Littlefield Publishing Group, Inc.
4501 Forbes Boulevard, Suite 200, Lanham, Maryland 20706
www.rowman.com

Unit A, Whitacre Mews, 26–34 Stannary Street, London SE11 4AB

British Library Cataloguing in Publication Information Available

Library of Congress Cataloging-in-Publication Data Available

ISBN 978-1-61146-175-6 (cloth : alk. paper) — ISBN 978-1-61146-176-3 (electronic)

∞™ The paper used in this publication meets the minimum requirements of American National Standard for Information Sciences—Permanence of Paper for Printed Library Materials, ANSI/NISO Z39.48-1992.

Printed in the United States of America

We dedicate this book to the Barber, Brann, Hanlon, Hogan, Lillis, Keil, and Mackin families who worked hard to carve out meaningful lives in Pennsylvania's anthracite coal region.

Contents

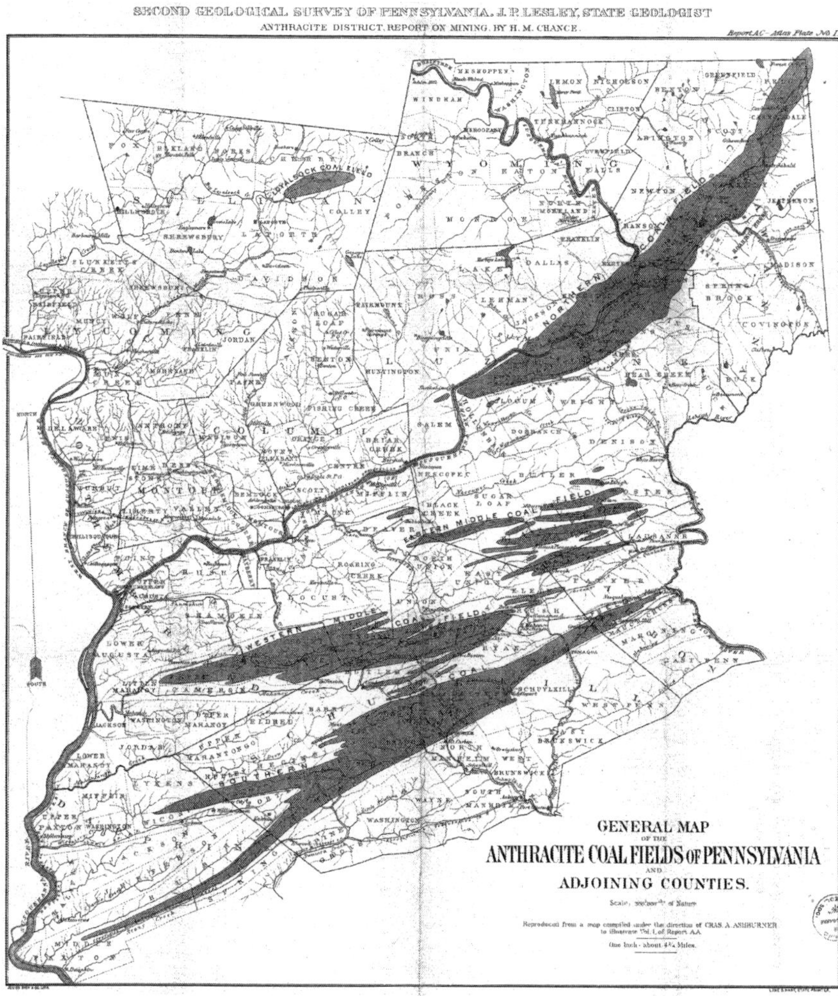

Map of Anthracite Coal Fields

Acknowledgments

We thank the external readers for their thorough analysis of the manuscript and their recommendations for change. We thank Amanda Fortenova, Archivist/Librarian, at the Luzerne County Historical Society. We also thank Margaret A. Hogan for the extraordinary job she did in copyediting the manuscript's final draft. Finally we acknowledge the outstanding work of Professor Michael M. Mirabito, Marywood University, Scranton, Pennsylvania, in contributing to our accessing historical photographs and for his original photographs that appear in this work

Preface

The goal of this work is to examine the anthracite coal trade's legacy in the five counties of northeastern Pennsylvania that constituted the core of this industry. Specifically, we examine the formation of the anthracite industry and its development from petty entrepreneurial production to production and distribution grounded in oligopolistic corporate enterprises operating in Carbon, Lackawanna, Luzerne, Northumberland, and Schuylkill Counties.

Chapter 1 presents an overview of the early history of what would become the anthracite coalfields and includes a discussion of the early to mid-nineteenth century split in the modes of production in the two sub-regions the northern fields made up of Luzerne and later, after their split, Luzerne and Lackawanna Counties, and the southern/central fields, made up of Carbon, Northumberland, and Schuylkill Counties with the former grounded in corporate production and the latter in entrepreneurial production. This split in the modes of production was based on the physical ecology of the two regions. This book argues that the initial conditions in which the anthracite industry developed led to differences in worker modes of organization and protest, and to the ways in which the two sub-regions were affected by the decline of the coal industry and the subsequent two waves of deindustrialization.

Chapter 2 discusses bourgeois class formation in the two sub-regions. It considers how class formation was a function of the nature of the local upper class both before and after the period of coal production and afterward as well as the nature and degree of urbanization in the two sub-regions, a difference that persisted into the twenty-first century and affected responses to the collapse of the coal industry and later substitute industries.

Chapter 3 deals with struggles between coal operators and corporations, on one side, and mineworkers on the other. It traces out types of pre-political rebellion by workers through various attempts to form coherent trade and, later, industrial unions. It also addresses the obstacles to union formation such as employer resistance, uneven conditions in the two major sub-regions, interethnic rivalries, and the like. In Chapter 4, we show how the workers' struggle against the mine operators continued, culminating in the success of the United Mineworkers of America (UMWA) as an organizing force. During the period covered in this chapter, difficulties in organizing mine-workers occurred due to splits between the fields caused by unequal working conditions, divergent political philosophies among mineworkers, the emergence of new ethnic cleavages as immigrants moved into the region from southern, central, and eastern Europe, and private and public repression by mine owners and the state. Nonetheless these factors did not suppress worker militancy, which grew out of the economic, political, and social exploitation that mineworkers were forced to endure as conditions of everyday life.

Chapter 5 focuses on the processes and events leading up to the 1902 strike, which secured a de facto place for the UMWA in the anthracite industry. The relative success of the 1902 strike was the result of the UMWA being able to suppress interethnic rivalries for the greater good of the workforce, the growing standardization of the industry's employment conditions due to the control exerted by J. P. Morgan's interests over the various producers, and the willingness of workers to set aside differences in political philosophies in pursuit of larger collective ends.

Chapter 6 demonstrates that the UMWA was not able to end pre-political forms of protest. Mineworkers often fought their own union and carried out so-called wildcat labor actions. After 1902, mineworkers also set up competing unions when they were not satisfied with the actions of the UMWA in cooperating with employers. The tension between mineworkers and the union was inevitable once the union became responsible for policing the workforce in the name of maintaining the terms of the contracts, which some segments of the workforce saw as more onerous than beneficial to the rank-and-file mineworkers.

Chapter 7 focuses on the second and third wave of deindustrialization and the consequences they have had in the two major sub-regions of the core anthracite fields, as well as the factors that brought about the decline in industrial production and its consequences. Chapter 8 looks at the contemporary post-industrial era and places it in the context of overall change in the anthracite region.

In discussing the patterns of economic organization and change, we rely on existing histories of the region. Neither of us are professional historians. We do not portray this work as discovering new historical facts. Our goal is far more modest. It is to use existing histories in order to weave a narrative of economic development and change that takes us from the early days of the region to the onset of the Great Recession, showing how the dismantling of the anthracite region's industrial infrastructure proceeded at a rapid pace even before the recession hit, replicating in some respects what had happened in the 1920s before the beginning of the Great Depression when the mining industry went into crisis and thousands of jobs were wiped out.

We focus on what has happened, and what is happening, since the decline and ultimate demise of the anthracite industry. We identify the replacement manufacturing industries, mainly the garment industry that moved into the region in the 1950s and 1960s, their subsequent declines, particularly the state of and labor struggles within the garment industry. We also examine the changes in patterns of employment in the region from 1997 to 2012.

Our study includes an analysis of the various political efforts at state and local levels to deal with industrial decline, and the regional variations in the success and failures of economic development policies in the five counties. Economic development policies have been ineffective in staving off industrial decline in both the northern and central/southern parts of the former anthracite coalfields. Both regions have seen an expansion of other, generally low-wage employment sectors. In most of these there have been significant increases in pay between 1997 and 2007, but the growth in wages has not been enough to offset the losses in income stemming from the dismantling of the industrial sector throughout the former anthracite coal regions. Given the pattern of differential changes in the former anthracite fields, we also show that the former coal region has somewhat bifurcated into two rather unevenly developed economic zones, with the northern region faring better than the central/southern region, a continuation of a pattern first established in the nineteenth century.

Chapter 1

A Brief History of Northeastern Pennsylvania's Early Settlement and Development

The northeastern Pennsylvania region, once known as "St. Anthony's Wilderness," due to its remoteness and wild terrain, has a long history of violent disputes over ownership and control of the region. The area was one of the centers of conflict between British forces, allied with Native Americans, and American rebels during the Revolutionary War. Around the same time, territorial claims exercised over the Wyoming Valley, an area about 110 miles west of New York and 90 miles northwest of Philadelphia, by the Susquehanna Company of Connecticut led to two wars between Connecticut settlers and ones loyal to Pennsylvania.

These Yankee-Pennamite Wars, taking place between 1769 and 1799, had seventeenth-century roots. The Connecticut Colony in northeastern Pennsylvania emerged as an outgrowth of Massachusetts after it received a royal charter from King Charles II of England in 1662, granting settlers the rights to lands that extended to the "South Sea." Later, in 1681, King Charles II granted territory, bounded on the east by the Delaware River and on the north by the 43rd parallel, which later became the anthracite coal center of Pennsylvania, to William Penn.

Following the Revolutionary War and the settlement of the dispute between Connecticut and Pennsylvania settlers, northeastern Pennsylvania began developing as a prosperous agricultural, proto-manufacturing, and trade center, especially around Wilkes-Barre. The principle manufacturing product in the region was textiles. By 1810 there were 250 looms in the Wyoming Valley producing textiles for export. The region also became a center of whiskey production, with a good number of distilleries developing in the early nineteenth century, as well as fishing (Hanlon and Zbiek 2003, 42, 43–46).

During the early years following the revolution, the people in the region began to create institutions to accommodate the needs of new populations streaming into the area, including migrants from other parts of Pennsylvania and European peasants displaced by the rapid capitalization of agriculture in the British Isles and Northern and Western Europe, especially the German states. In 1807, the emerging manufacturing, export agriculture, merchant, and professional elites established the Wilkes-Barre Academy to prepare their sons for college. No comparable institution was built for the children of laborers or small farmers. The opening of the school further strengthened the division of the region's population along social class lines, with an educated elite on top, a semi-literate agricultural and industrial workforce at the bottom, and a thin middle class separating the two.

By this time, Wilkes-Barre was developing into the commercial, wholesale, financial, and professional services center for a vast agricultural hinterland that stretched north to the New York border and almost to Easton in the south, making Wilkes-Barre the most significant city in the northeastern part of the state. Later Wilkes-Barre's preeminent position as the leading city in the region would be usurped by Scranton, a city to the north of Wilkes-Barre, which would become the seat and population center of Lackawanna County, formed out of Luzerne County in 1878.

Carbon County had a somewhat different history. The land, in what would become Carbon County, was sparsely settled by Native Americans before its colonization by whites. Under the leadership of Count Nicolaus Zinzendorf, Moravians began large-scale European colonization of the region in 1746. It was initially a part of Bucks County, under the control of Pennsylvania (Campion 1997). Coal was discovered in the area in 1786, which had a transformative effect on the economy, society, politics, and ecology of the area (ibid.).

The land that would become Northumberland County was acquired from Native American tribes through a series of purchases. William Penn bought a large tract of land in 1682, followed by a number of other transactions, which eventually gave Penn control of the lands between the Delaware and Susquehanna rivers (Bell 1891, 42–59). There was contention among the Native American tribes as to the validity of the agreements with Penn, and violence erupted on the frontier in 1755, with Native Americans and Pennsylvania settlers fighting each other. Furthermore, Native Americans allied themselves with the French in their struggles against English colonists, and violence between Native peoples and colonists continued during the Revolutionary War. After the revolution, agriculture was the dominant industry in Northumberland County until it eventually was displaced by coal mining and railroad operations.

Even before Europeans had settled in the region, Native Americans living in various parts of northeastern Pennsylvania, including the Wyoming Valley, had been aware of the presence of anthracite coal and some of its uses. This regional natural resource also had desirable properties for the European settlers, such as its ability to burn cleanly, and therefore appeared to have possibilities for heating homes and serving as an industrial fuel in blacksmith shops and the like. The major problem with the fuel was that it would not burn consistently without a continuing draft.

While the earliest recorded use of anthracite coal by white settlers was in 1769, it was not in high demand until 1808 (Goin and Raymond 2001). That year, Judge Jesse Fell, a local hotelkeeper and politician in Wilkes-Barre, discovered that he could obtain a consistent burning of coal by breaking it into small pieces and burning it on an open grate. The discovery of the open grate method of burning anthracite made the fuel a popular source for industrial purposes and an alternative to wood burning for heating homes.

Even though the open grate first appeared in Wilkes-Barre, Luzerne County was not the earliest center of anthracite mining. Instead, the mass production of anthracite coal began much earlier in coal beds to the south of the county. Colonel Jacob Weiss, a Revolutionary War veteran, was using anthracite as early as 1786 in what would become Carbon County. In 1787, Phillip Ginder discovered a massive vein of coal, later called the Mammoth Vein, on Sharp Mountain, near Pottsville in Schuylkill County. Ginder's discovery led to the formation of the Lehigh Coal Mine Company and, eventually, to the establishment of the powerful Lehigh Coal and Navigation Company (Campion 1997). A significant pioneer in the coal business in Carbon County was Asa Packer. Born in Mystic, Connecticut, he found his way to Mauch Chunk, in what was to become part of a newly formed Carbon County. There he became involved in transportation, leasing of coal lands, and mining of coal. Packer was involved in the development of the Lehigh Valley Railroad in 1852. By this time, Packer already was a wealthy entrepreneur and politician, having served in the Pennsylvania legislature; while there, he played a key role in separating Carbon County from Northampton County. He also served a five-year term as a county judge and, between 1852 and 1856, he served two terms in the U.S. House of Representatives. Packer was a zealous opponent of workers' rights to organize and strike. He had tried to break up the boatmen's strike of 1843, one of the earliest worker strikes in the region, only to find himself pitched into a canal by the workers for his troubles (Campion 1997, 34).

Coal initially was found in what would become Northumberland County in 1780. The first practical use of anthracite coal from Northumberland County took place in 1810 when it was used on a regular basis by a blacksmith and farmer named Thomlinson (John 1891, 359). Anthracite coal production was stimulated by the War of 1812, when Pennsylvania iron mongers were cut off from their supplies of Virginia coal.

John Thompson, a fifteen-year-old boy, took the first coal to market from the settlement of Shamokin in 1814. He carried the coal from his father's farm to Sunbury. Mining in its early years in and around Shamokin was hit or miss, with many would-be operators not finding productive veins. As a result many operations opened and closed in a short time. It was not until the early 1820s that mining began to operate on a somewhat sustained basis. In its early years, United States' mining was a simple operation in which unskilled men dug veins that had been exposed by running water.

Between 1824 and 1829, mining led to wild land speculation, with investors hoping to capitalize on the discovery of coal seams. One of the largest, Stephen Girard of Philadelphia, worked with local investors to start The Danville and Pottsville Railroad.

Anthracite production was stimulated by the formation of the Schuylkill Navigation Company in 1815 and the Lehigh Coal and Navigation Company in 1821. In 1832, the Delaware Canal opened. It facilitated bringing anthracite to Philadelphia and New York. Following these early canals a network of canals and, eventually, railroads spanned the anthracite region.

In 1835, the town of Shamokin was established. About this time, drift mines and slope mines began opening throughout the area. But it was not until 1839 that the coal trade finally blossomed in this region, shipping 11,930 tons over the year (ibid., 365).

By 1842, the Shamokin area was caught up in the national economic crisis of that year and a large number of people began leaving the region. Mines closed and railroads went bankrupt and shut down operations. Around 1850, matters began to turn around and a modicum of prosperity returned to the area. A number of new coal operations began and railroads re-opened, sometimes under new names with different owners. By 1853, shipments of anthracite had increased to 15,500 tons a year, a small increase over 1839. In response to production increases the 1850s saw the opening of four new coal breakers.

Coal also began to be mined and processed in large amounts in Mount Carmel around 1854. The tonnage being shipped from the Northumberland County area had grown to 19,640 tons by that year. The European and American financial crisis of 1857 seriously affected the county. One of its biggest railroads, the Philadelphia and Sunbury Railroad, a major hauler, together with all of its equipment, coal lands, and leases to operators were sold at a sheriff's sale. Coal production fell in 1858 as a result of the financial panic.

Coal prices increased dramatically during the Civil War, as did production in Northumberland County. By late 1864, prices began dropping, as demand fell while production increased and mineworkers' wages were reduced significantly, much to the anger of the workers. By 1868, Northumberland County was shipping 911,784 tons of anthracite (ibid., 381). In 1869, mineworkers in Northumberland were organized into the Workingmen's Benevolent Association (WBA) and participated in the industry-wide strike of that year.

In 1871, the coal industry underwent a dramatic structural change in Northumberland County and Schuylkill County. While producers in the southern and central fields faced stiff competition internally, competition also was severe between the central and southern fields and the mine operations in the Wyoming and Lackawanna Valleys, where giant corporations owned by northern region investors and capitalists from outside of the region mined and transported coal and were able to regulate competition and production among themselves. In response to railroads in the north becoming integrated with production, the three largest railroads operating in Northumberland County organized subsidiaries to become directly engaged in mining operations of one kind or another. For example, the Reading Railroad formed the Philadelphia and Reading Coal and Iron Company, a wholly owned subsidiary of the railroad. During 1871, this new company purchased vast tracts of prime coal land in Northumberland and Schuylkill Counties. All of this was financed by an $11,000,000 loan from Philadelphia's Fidelity Trust and Safe Deposit Company (ibid., 383). The loan was secured by stock from the railroad. Despite the increasing consolidation of production in Northumberland County, there were still a large number of small, independent production operations remaining in the 1870s, creating instability in the central and southern fields that was generally lacking in the mining operations to the north in the Wyoming and Lackawanna Valleys.

Sunbury, the county seat of Northumberland County, while more than doubling in size from 1860 to 1890 (growing from 1,869 to 5,930), did not occupy the preeminent positions that Mauch Chunk held in Carbon County, Scranton in Lackawanna County after 1878, and Wilkes-Barre in Luzerne County. Sunbury's population in 1890 was surpassed by Shamokin Borough (14,403), the largest settlement in the county; Coal Township (8,616); and Mount Carmel Borough (8,254) (Bell 1891, 154).

Anthracite coal was discovered by accident in Schuylkill County in 1790 by an itinerant lumberman from Broad Mountain named Nicho Allen. Colonel George Shoemaker tried to stimulate a market for anthracite by shipping a load to Philadelphia from Schuylkill County in 1812, but his attempt met with limited success (Luther 1881, 42). Not until 1822 did coal begin to be shipped by the Schuylkill Navigation Company's early, rudimentary canal, establishing a market for anthracite from Schuylkill County.

In 1829, the Schuylkill region was beset by wild land speculation, as would-be millionaires flocked to the region seeking their fortune by securing coal leases. Most ended up with little or nothing in return for their investments and their labors. For those few who understood the geology of the anthracite deposits fortunes were made as land prices escalated.

Between 1825 and 1829, coal production in Schuylkill County increased continually and operators received good prices for each ton shipped. They had every reason to expect such trends to continue, but things did not work out that way. There were serious problems of overproduction in 1831 resulting in falling prices for anthracite and significant wage reductions for skilled miners and their laborers. This was but an early instance in what would become a long history of overproduction crises in the industry (ibid., 47).

In order to try to regulate the industry, in 1832 major colliery operators and other actors connected to mining founded the Coal Mining Association of Schuylkill County (ibid., 47). In its first report, the Association generated an estimate of the total value of capital invested in the Schuylkill County anthracite industry. It found that the costs for railroads "as about $656,000; the amount invested in coal lands and buildings in the county was estimated at $600,000; the amount expended in opening veins of coal, in building fixtures, cars, etc., connected with mining was $200,000, to which may be added 500 boats, averaging $500 each, for a total of $250,000; and the total investment is shown to have been $7,106,000." Furthermore according to the report, "Not an individual miner engaged in its commencement was supposed to have realized a cent of profit" (ibid., 48).

In Schuylkill County, there was strong opposition to the formation of coal corporations in the early 1830s. The belief among a large proportion of the county's residents was that mining was and should continue to be an individual, entrepreneurial activity that required little in the way of capital investment. The population was especially opposed to combining mining with railroads and canal companies, fearing that such conglomerates would effectively wipe out petty producers and thereby change the nature of the industry in the county. A Pennsylvania legislative committee investigated the issue of corporate mining and decided to continue restricting corporations from operating in Schuylkill County.

In the early days in the Schuylkill field, mining was a relatively simple activity. The mines were one- or two-man operations, where coal was dug from outcroppings on the surface of lands. Coal diggers would expose a seam as far as they could, then put wet lime on the seam. This would crack the coal as it dried, and the miners would shovel the coal into a wagon, from where it would be taken to a local railhead and then on to the Schuylkill

Canal. The ease of entry into the business drew many migrants into the coalfields from rural areas surrounding the production points in Pottsville and the rest of the Schuylkill. In Pottsville alone in 1833, there were at least fifty operators working such mining sites and between then and 1870, when the Philadelphia and Reading Coal and Iron Company began to monopolize production in the region, there were over a 1,000 independent operators who had worked or were working in the Schuylkill field (Davies 1985, 107). Most of these operators worked on leased land. It was in the interests of the landowners to encourage as many operators to work their lands as intensively as possible, thereby maximizing their profits, but, in turn, this policy created marginal mining operations that were in intense competition with each other.

By 1835, drift and slope mining were beginning to appear in Schuylkill County, allowing for limited mining below the water table (Luther 1881, 48). These, like the earlier mining methods, were low technology operations (Yearly 1961). Drift mining was a simple operation in which a tunnel was dug into a mountain coal seam horizontally and then the coal seams mined vertically. The water ran down the incline and out of the mine. Slope mines required a bit more technology in that they involved digging a tunnel down into a seam and using pumps and more labor to remove the water. In neither case were these costly endeavors.

The financial panic of 1834 caused a substantial drop in demand for anthracite, which, among other things, led to massive unemployment in the industry and a large oversupply of coal for the following season. By 1836, demand increased and the oversupply had been cleared to the extent that coal shortages were appearing (Luther 1881, 50). In the face of these shortages, the U.S. Congress began considering doing away with the tariffs on foreign coal. To show that Congress was wrong and that domestic producers could meet demand, corporate and individual coal operators greatly accelerated production. In hindsight, this turned out to be a disastrous decision, as another financial panic began in 1837, once again reducing demand. The coal producers were stuck with a huge oversupply as demand plummeted, operators responded by suspending operations at a number of production points. Fortunately, the panic was short-lived, and demand increased again and coal shipments once again increased to meet the need for anthracite. The situation in the coalfields was complicated by the onset of a depression in 1838, demonstrating once again the instability that would become a major characteristic of the industry.

The early mines in the central and southern fields were often part-time owner operations run by local farmers or village artisans, and typically had only small workforces digging coal. Coal extracted from these mines was shipped southward using rivers and roads, and eventually canals and later railroads, to Philadelphia, Baltimore, points farther south, and, via the ports of Philadelphia and Baltimore, to New York City and Boston markets.

As American cities grew and a manufacturing infrastructure was built, it soon became evident that the anthracite fields to the south of Luzerne County were not going to be able to meet the country's energy needs. Exploration began to take place in possible coal locations farther to the north in Luzerne County, where there were already scattered operations. At this point, however, it was not clear how much coal was available or how accessible it might be in what would become the Wyoming and Lackawanna fields.

Beginning in the 1820s, Luzerne County coal was already being shipped from around Carbondale—which at the time was in Luzerne County, as was the entire Lackawanna field—to New York on a railroad line developed due to financing by New York capital, the Delaware and Hudson Railroad (D & H). This rail line ran from Carbondale, near the northeastern boundary of the anthracite coalfields, to Honesdale on the Lackawaxen River. In 1829, the D & H shipped its first loads of coal from Carbondale to New York City, opening up what would become a flourishing trade between New York City and the northern anthracite fields.

By the early 1840s, geologists had determined that the northern field potentially had greater reserves than what were known to exist in the anthracite fields to the south. Geologists and other experts, however, noted several features of the region that would make coal extraction difficult. First, the coal lay deep underground and would require mining technologies far more complex than mere slope or drift mining to fully exploit deposits. Second, it was going to be difficult getting the coal to market given the general inaccessibility of the entire region, which was mountainous and heavily forested. Third, there was not sufficient local entrepreneurial capital available to open these deep mines, and fourth, there was an inadequate supply of labor with the necessary skills to work in deep mines.

Given the expense of opening and operating deep mines, the only way to develop the Wyoming and Lackawanna fields was for the local bourgeoisie to gain access to non-local capital by forming corporations, which drew on local and extra-local sources of capital. This pushed out the small operators who had begun the industry around 1800 in the northern regions. Mining companies required such high levels of capitalization that could only be found in the hands of the local bourgeoisie and outside investors. Capital demands were too much to bear for the local bourgeoisie to manage alone.

Capitalists from New York, Philadelphia, Baltimore, and Great Britain came together with local capitalists to invest in opening the Wyoming and Lackawanna fields and developing a transportation system for exporting coal from the Wyoming and Lackawanna Valleys. Therefore, almost from its very start, the northern anthracite region was dominated by corporate capital, rather than personal, entrepreneurial capital, as was the case in the anthracite fields to the south. As a result of its corporate development, the Wyoming field and its northern geological extension centered on Scranton, the Lackawanna field, were vastly out-producing the southern fields by the 1860s. One of the factors that led to greater production in the north was not merely greater coal reserves, but the presence of a strong oligopolistic cartel formed among the major corporations in the north that could control levels of output and prices (Keil 1982, 1988).

At this time, many operators in the central and southern fields spent less than $500 in capital to open and operate their mines and there were a large number of lessees that had even less capital investment. It required a smaller amount of capital to open and run a mine than it did to buy a decent-sized farm and stock it with animals (Luther 1881, 48). Because of low entry and low operating costs there was a continuing over supply of operators in the central and southern fields. This resulted in a continuing anarchy of production marked by intense competition to produce ever more coal and intense price competition. This anarchy of production lasted until after the end of the Civil War. At this point political changes were taking place in Pennsylvania politics that would have implications for issues like granting corporate charters in the coal industry. During the 1830s, such changes included the increased significance of professional lobbying on behalf of business interests, escalating corruption that would become endemic by the late 1840s, and unprecedented favorable treatment for businesses in the legislature, reflected in an increase to 420 new corporate charters granted in the 1830s from 144 in the 1820s (Bowers 1983).

As early as the 1820s, the Pennsylvania legislature allowed the Lehigh Coal and Navigation Company (in 1822) and the Delaware and Hudson Company (in 1824) to incorporate and granted charters known as double-privilege charters that allowed them to legally own, mine, and transport coal for the Central and Wyoming fields, respectively. Incorporation and double-privilege charters remained legally prohibited for the Schuylkill field. In March 1833, the state legislature of Pennsylvania created a commission, chaired by state senator Samuel J. Packer, to look into the effects that further incorporation of coal companies would have on the coal trade. The committee rejected an increase in the incorporation of coal companies, especially when they were combined with canal and railroad companies (Luther 1881; Wardell and Johnston 1987). This decision perpetuated decentralized, entrepreneurial coal production in the anthracite fields. Later, general charters became legal in Pennsylvania in 1849, and an 1853 supplement to this legislation (passed in 1854) made the coal industry eligible, thereby leading to the lifting of general bans on coal company incorporation in Luzerne, Northumberland, and Carbon Counties, but not Schuylkill (Adams 1998). The petty producers had enough political influence in Schuylkill County to block the legislation from applying there. A second general corporate act, passed in the 1854 state legislature, was extended to Schuylkill County in 1857, thereby laying the foundation for consolidation of the anthracite industry there.

Beginning in the 1840s, northern and central/southern anthracite regions experienced rapid increases in the demand for coal and consequently labor. In 1830, the United States was still relying on small-scale production and imports to meet demands for iron, but during the 1830s and 1840s, the situation began to quickly change (Chandler 1972). In the 1830s, with the cost of wood increasing due to the deforestation of the Northeast, anthracite was used to power industries that previously had depended on wood. After 1840, anthracite coal was key for the smelting of mass-produced iron where fuel costs mattered (Marsh 1987). The use of coal in the domestic fuel-intensive iron production process has been identified as a key event in American economic history (Chandler 1972).

Anthracite coal areas experienced population booms due to the growth of mining and allied industries. For example, Wyoming Valley's urban base grew by 2,000 percent between 1840 and 1880 (Davies 1985). Table 1.1 reports Census population figures for the Pennsylvania anthracite coal counties of Luzerne, Lackawanna, Carbon, Northumberland, and Schuylkill Counties between 1850 and 1920. High growth rates took place in all anthracite coal counties, thereby creating the need for real estate development, housing construction and building materials, road and bridge construction, local utility construction (including sewer and water disposal systems), bank development, wholesale and retail trade, and growth in the professions. All of these provided profit opportunities for local capital.

Table 1.1 Population Totals for Pennsylvania Anthracite Counties, 1850–1920

	1850	*1860*	*1870*	*1880*	*1890*	*1900*	*1910*	*1920*
Northern Field								
Luzerne & Lackawanna*	56,072	90,244	160,915	222,334	343,291	450,952	602,756	677,302
Luzerne				113,065	201,203	257,121	343,186	390,991
Lackawanna				89,269	142,088	193,831	259,570	286,311
Central/Southern Fields								
Carbon	15,686	21,033	28,144	31,923	38,624	44,510	52,846	62,565
Northumberland	23,272	28,922	41,444	53,123	74,698	90,911	111,420	112,079
Schuylkill	60,713	89,510	116,428	129,974	154,163	172,927	207,894	217,754

*In 1850–1870, the U.S. Census reported data for Luzerne County, comprised of what today is known as Luzerne and Lackawanna Counties. Lackawanna County was formed from Luzern County in 1878, so Lackawanna County does not appear until the 1880 Census.

Source: U.S. Census of Populations and Housing. Available at http://fisher.lib.virginia.edu/collections/stats/histcensus/index.html. Accessed July 18, 2007.

When coal mining was a relatively simple operation that did not need complex skills, laborers, as previously noted, were drawn from the immediate countryside and the small towns that dotted the regions. As deep mining developed, however, labor demands increased, as did the skill levels necessary to extract coal. Foreign labor began streaming into the coal regions. Initially, a good deal of the immigrant labor came from the British Isles, especially from England and Wales, with smaller numbers from Scotland and Cornwall. They made up a stratum of skilled labor. In the 1840s, Irish immigrants fleeing the Great Famine began moving into the coal regions. They took unskilled jobs, which put them in competition with native labor and, in the southern and central fields, with Pennsylvania Germans, whose origins were mainly agricultural. Miners from Germany also immigrated into the region. The heterogeneous nature of the labor force would be a continuing problem for those trying to organize the workforce into a coherent labor movement, as shown in later chapters. The organization of workers was also complicated by a large pool of surplus workers, a reserve army of labor who could be drawn into the labor market during times of strikes and increased demand. Furthermore, the chances for worker organization were limited by private and state repression.

These large-scale economic, social, and demographic changes led to changes in the class structure and class relations in the region. In addition, the anthracite industry has been one of economic uncertainty and economic turbulence. This instability contributed to making the work capricious, especially in the late nineteenth century (Phelan 1994). These conditions contributed to social disruption, worker insecurity, and outright conflict. In the next chapter, we examine the class structures that emerged from the formation of the mining industry in northeastern Pennsylvania and the nature of class conflict that permeated the region. The division of the anthracite mining industry for a good part of the nineteenth century into an entrepreneurial sub-region in the southern and central fields, and a corporate region in the Wyoming and Lackawanna fields, has had a continuing effect on northeastern Pennsylvania's economy, with the southern and central fields lagging behind the development of the Wyoming and Lackawanna fields. The entrepreneurial nature of the southern and central fields contributed to the boom-and-bust cycles that characterized production for much of the nineteenth century. The intense competition led to continuing problems of overproduction and price-cutting, which put downward pressures on workers' wages and working conditions in the mines. Moreover, the different conditions of production produced different work experiences for the men in the industry, making it difficult for them to coordinate action across fields.

Chapter 2

Bourgeois Class Formation in the Coalfields

The central and southern anthracite fields never developed cities that played dominant roles comparable to Scranton in Lackawanna County or Wilkes-Barre in Luzerne County, with the possible exception of Pottsville before 1870. Instead, the central and southern coalfields were far less urbanized and continue to be so today. Luzerne and Lackawanna Counties are more urbanized, with a scattering of smaller cities distributed throughout, such as Carbondale and, later, Scranton in Lackawanna County; Pittston, Nanticoke, and the largest city Wilkes-Barre and Hazleton in Luzerne County. In the central and southern fields, there are Pottsville in Schuylkill County; and Shamokin and Sunbury in Northumberland County. In the absence of significant urbanization, the central and southern fields, dominated by petty producers until the Philadelphia and Reading Coal and Iron Company moved into the region as a major producer after the Civil War, never developed a substantial bourgeoisie that was oriented to urban development on a large scale or played a significant role in the development of the coal industry after the 1870s. In Wilkes-Barre and later Scranton, the formation of a local bourgeoisie strengthened the development of somewhat more diverse economies, and their investments stimulated urban development.

Pottsville, Schuylkill County's seat, developed a bourgeoisie that never successfully transformed itself into a hegemonic class that could dominate the southern anthracite coalfields in the same way that the Wilkes-Barre and Scranton bourgeoisie did in their respective regions (Davies 1985). In particular, after 1870, Pottsville's role in the Schuylkill field began to decline in the face of inter-urban competition.

The boom in the demand for anthracite in the 1820s created the conditions for Pottsville's establishment and growth over the next fifty years. The settlement's businesses provided goods and services to the growing number of small-scale mine operators that were flooding the southern coalfields. At the time, Pottsville's main rival was the incorporated town of Mount Carbon, which had the advantages of being the oldest settlement in the region and the terminus of the Schuylkill Canal. Mount Carbon lost out in the competition for dominance in the Schuylkill region, however, despite its initial advantages.

In 1828, having experienced growth and development, Pottsville petitioned the state legislature for incorporation, thereby positioning itself as a direct rival of Mount Carbon. In the face of opposition from Mount Carbon, the state's leaders took three years to approve Pottsville's request for incorporation. But Pottsville's capitalists were not satisfied merely with their political victory. In alliance with capitalists from Philadelphia, they succeeded in extending the Schuylkill Canal to Pottsville (ibid., 104). As a result, Pottsville emerged as the dominant urban center in the southern coalfields, far surpassing Mount Carbon. As Pottsville prospered, it drew migrants from surrounding rural areas and from Europe, principally Germany, England, Wales, and, later, Ireland.

But given the dispersed, entrepreneurial nature of mining in Schuylkill County, elements of the bourgeoisie remained fragmented and never formed a coherent, class-conscious bourgeoisie. Some of the coal entrepreneurs did reinvest a portion of their profits in Pottsville and built institutions commonly associated with the upper class, such as private clubs. Others, however, seeing residence in Pottsville as a temporary sojourn, invested in Philadelphia and other metropolitan areas, and eventually migrated from Pottsville, abandoning the city for what they saw as better social, business and financial opportunities elsewhere. There also was a significant out-migration of children of the bourgeoisie, who saw their economic opportunities shrink with the invasion of Philadelphia corporations after 1870 and the growing rivalry with anthracite producers in the Wyoming and Lackawanna Valleys.

The subordination of the Schuylkill field to Philadelphia financial interests deprived the region of the financial and human capital that would have been necessary to create an economic infrastructure to sustain urban development such as was taking place in the Luzerne and Lackawanna fields. It also lowered businesses' social commitment to the area. In this sense, they were more like the elites in Mauch Chunk in Carbon County, some of whom abandoned their city to develop their own institutions in South Bethlehem (Folsom 1981; Campion 1997).

Pottsville's role in the anthracite industry was not only diminished by this out-migration but also by the opening of mines farther north in Schuylkill and Northumberland Counties. The growth of production and small settlements in the northern portions of Schuylkill County by 1890 led to over half the population living in the northern rim of the county. Pottsville had little influence over the coal industry in this area and, in fact, only controlled about 25 percent of the industry, overall in Schuylkill County, at the peak of its influence (Davies 1985, 113). After 1870, outside interests came to control production and ownership of coal lands throughout Schuylkill, further weakening the impact of Pottsville's bourgeoisie in the industry.

Pottsville's bourgeoisie was further weakened by the depression of the 1870s, when a number of the banks it controlled failed and its major businesses went into bankruptcy. The depression, along with corporate penetration, resulted in a completely different configuration of local power than what existed before corporate penetration and the 1870 downturn. The Pottsville bourgeoisie's power over other communities in the county was seriously diminished. It no longer was the center of financial power in its county. The historian Edward Davies provides a detailed account of bank failures during the 1870s and the impact these had on all strata of the community, including the bourgeoisie themselves, who saw substantial losses of personal savings and business capital (ibid., 121–23).

The economic ravages of the 1870s put considerable strain on a bourgeoisie that shared little in the way of mature social and political ties to bind it together. Unlike the case in Wilkes-Barre and, to a lesser degree, Scranton in Lackawanna County, where the bourgeoisie was highly homogeneous with respect to religion, ethnicity, social class of origin, and family background, Pottsville's bourgeoisie was divided on all of these dimensions, so it never achieved the class coherence that took place in the cities in Luzerne and Lackawanna Counties.

Davies notes that in Pottsville, "Boom-town growth had produced an extremely heterogeneous elite, whose members shared only entrepreneurial and occupational experiences" (ibid., 128). During the economic crisis of the 1870s and the consequent slowing of growth in the 1880s, these economic ties proved insufficient in maintaining class coherence and solidarity.

By 1870, the Pottsville bourgeoisie had begun to overlap considerably with the political elite in Schuylkill County and was in the process of forming itself into a closed-off social aristocracy: an urban bourgeoisie. The bourgeoisie was being formed into a powerful bloc that was increasingly segregating itself residentially and in other ways from the general population. None of this, however, matched what had taken place in terms of upper-class integration and political dominance in Wilkes-Barre and Scranton.

The men in the Pottsville bourgeoisie also were in the process of building social and economic ties throughout the southern rim of Schuylkill County in the years before 1870. By 1870, these ties had multiplied considerably. Again, the 1870s depression and penetration of external capital weakened the integration of Pottsville's bourgeoisie with other urban elites in the southern rim. These other elites no longer had to be dependent on Pottsville's bourgeoisie for the marketing of coal or as a source of business capital. Given the lack of social integration among the various elites in the region, the lessening of economic dependence tended to strain, if not completely break, the ties that existed among the various regional elites in the southern rim, where those ties had been the strongest. Connections among the bourgeoisie in the northern rim and in Northumberland County with those in Pottsville were never as strong as ties in the southern rim between elites from Pottsville and other communities. After 1870, Pottsville's ties to the northern rim and Northumberland County had only marginal significance. Thus, after 1880, Pottsville lost whatever chance it had of integrating and subordinating the remainder of Schuylkill County and Northumberland County under its influence. It lost its possibility of becoming the preeminent urban community in the southern and central anthracite regions.

Before the onset of major coal production in Luzerne County in the 1850s and 1860s, the Wilkes-Barre region was dominated by a small circle of around 100 households, 95 percent of whom were related by marriage or by direct descent (ibid.). Part of this circle was descended from the original Connecticut settlers who had come into the region as farmers prior to the Revolutionary War, and who had fought against Pennsylvania settlers after the revolution, in the Yankee-Pennamite Wars. Even though Connecticut lost its political claims in Pennsylvania, the property owners who had bought land through the Connecticut-based Susquehanna Land Company were allowed to retain their personal titles of ownership. Many of these Connecticut settlers stayed on in the region and came to form part of the regional bourgeoisie, partly through their role as petty rentiers.

In 1849 and through the 1850s, a bourgeoisie of some substance emerged in Wilkes-Barre and Luzerne County. It was organized around banking, merchant capital, the law, the leasing of coal rights, and other professions. They remained the dominant local economic and political actors from the early part of the nineteenth century well into the early twentieth century in ways not seen in other parts of the coalfields, except for Lackawanna County. Despite the rapid and large-scale economic and demographic changes taking place between 1840 and 1930, there was a high degree of continuity in the Wilkes-Barre bourgeoisie. Eighty-six percent of the leading families effectively maintained their bourgeois standing through the 1920s (ibid.). This was not the case in Scranton, where the bourgeoisie, except notably the Scranton family, had difficulty perpetuating itself across generations (Folsom 1980).

A pig-iron manufacturer named William Henry opened an operation in the Lackawanna Valley in 1842. The brothers Seldon and George Scranton invested $20,000 in the operation. George was William Henry's son-in-law. The Scranton brothers, at the time, operated the Oxford Iron Works in Oxford New Jersey. The initial investors soon discovered that there was virtually no local market for their iron and there was no profitable way to get it to other markets. George took over the business and turned into a nail making operation. He persuaded two cousins, Erastus and Joseph Scranton, prosperous merchants in Augusta Georgia, to invest close to eighty thousand dollars in a rolling mill and nail factory. This business also failed because of the low quality of the nails.

The Scrantons next turned their hand to manufacturing railroad rails and, in 1848, they received a contract from the New York and Erie Railroad to manufacture rails for them. They managed to convince the railroad to advance them funds to build the rails. In 1853, they founded the Lackawanna Iron and Coal Company. Using capital supplied by New Yorkers, they also built two rail lines, which they later used to form the powerful Delaware, Lackawanna, and Western Railroad. In addition, they began opening and operating mines in the Lackawanna Valley (Miller and Sharpless 1985, 62–63).

In and around Wilkes-Barre, kinship, interfamily connections, and a solid network of class-based institutions held the various economic elites together and helped transform them into a true bourgeoisie. They had their own small number of churches, golf and other sports clubs, an historical society to honor their families' "achievements," their own private grade schools and a private high school, and the crown jewel, a private "gentlemen's club," the Westmoreland Club, the membership of which spanned the entire coal region and beyond. In its heyday, membership in the club was restricted to the region's male Protestant gentry. It now accepts women and non-Protestants as members, as well as blacks.

The various elites, by transforming themselves into a fully developed bourgeoisie, managed to exercise exclusive economic hegemony and political dominance of the region until the penetration and rapid expansion into the area by external corporate capital around the mid-nineteenth century. Corporate interests, in alliance with Wilkes-Barre's bourgeoisie, took control of the local coal industry in the Wyoming and Lackawanna Valleys and held their control over the coal industry for almost a century. With external corporate infiltration of the region, local capital lost some of its hegemonic power. It no longer had the exclusive capacity to develop the region with a view toward maximizing the economic, social, and political interests of local capital. Economic development of the region was now to be shared with extra-local corporate capitalists and their surrogates (Keil 1988).

The old bourgeois families, however, managed to maintain their rôle as a political ruling class. They continued to dominate local, state, and national elective and appointive offices based in the region and commanded local militias and the civil and criminal courts in Luzerne County. Between 1865 and 1875, members of the old-stock bourgeoisie in Wilkes-Barre held forty-two out of fifty-two possible elective offices in the city (Davies 1985). They also held most county elective positions and controlled local, state, and county judicial positions. At one point, politicians from Wilkes-Barre held the offices, concurrently, of governor, sitting member of the state supreme court, leadership positions in the legislature in both houses, and leading positions in the Democratic and Republican parties at the state level.

Even though the old bourgeoisie lost exclusive hegemony over the local economy, corporate investment created a new "subsidiary" range of profit opportunities for the local upper class in developing the physical and social infrastructure of mining. They built, for example, the basic infrastructure of utilities, including light rail lines that tied the anthracite cities together and, later, trolley and bus systems, a local system of banks, institutions for venture capital, the lumbering industry, textile mills, and other manufacturing concerns. They also controlled public institutions, especially those of repression.

Several members of the old local bourgeoisie served on statewide boards and commissions, including the Canal Commission, which authorized the building of a canal along the North Branch of the Susquehanna River, stretching from the New York State line to the Pennsylvania Main Canal. The canal not only greatly facilitated the movement of coal to Philadelphia but it also provided for the extensive and rapid expansion of other trade networks centered in and around Wilkes-Barre. Thus, even though a good part of the local bourgeoisie was reduced almost to the status of a regional comprador bourgeoisie, especially after 1880, it still retained enough economic and political independence for the accumulation of large fortunes in service of the mining companies and their interests, as well as its own local interests.

In the 1820s and 1830s, strong indications suggested that Carbondale was going to become the premier city in northeastern Pennsylvania's Lackawanna Valley (Folsom 1981, 20). First, it had the advantage of its close location to markets for coal in New York State and New York City, compared to other settlements in the region. Second, it had an initial advantage in shipping coal. Third, it had attracted corporate investment in the development of mines and mining. Fourth, it attracted labor from the surrounding area and Europe, allowing the city to grow at a rapid rate.

By the 1850s, however, Carbondale had lost out to Scranton as the first city of the Lackawanna Valley. Scranton would also eventually begin to eclipse the much older city of Wilkes-Barre as the principal settlement of the northern coalfields and of the anthracite region as a whole. Scranton's growth was predicated on the presence of an entrepreneurial class of migrants, many of them from New Jersey, like the Scranton family itself (ibid.). This proto-class was geared toward promoting the growth of the city in order to enhance its own economic prospects.

Scranton grew despite the active opposition of older settlers and Wilkes-Barre's bourgeoisie. The old settlers in Scranton differed in important ways from the new, entrepreneurial immigrants. These newer entrepreneurs and industrialists were younger and more educated. Many of the old settlers resented the intrusion of these new entrepreneurs into their region and their transformation of the region (ibid., 30, 32).

Wilkes-Barre's politically active bourgeoisie, working with the D&H, managed to block for several years Scranton's attempt to build a rail line to connect it to New York markets, fearing that the presence of a railroad would undermine Wilkes-Barre's hegemonic position over trade in the Susquehanna River basin. Wilkes-Barre's politicians also blocked Scranton businessmen's attempt to form a new county until 1878, when Lackawanna County was carved out of Luzerne County. Scranton secured its railroads when, between 1850 and 1853, two lines were constructed that eventually were merged into one—the Delaware, Lackawanna, and Western Railroad (DL&W). Financing was obtained in part from New York City capitalists who had close financial ties to the Erie Railroad (ibid., 33).

In addition to investments in railroads and iron works, the Scranton family, who dominated the class structure of the Lackawanna Valley, invested heavily in land acquisition and development, building houses for their workers and mansions for the local bourgeoisie. They also turned to coal mining. The DL&W's charter had given it mining rights, which it exercised to great profit, especially since the coal was hauled on a Scranton family railroad, which they owned in conjunction with New York capitalists. They also founded the Scranton Coal Company, which was financed in part by New York capital as well (ibid., 39).

Scranton's economy became ever more coal dependent, but, at the same time, there was some degree of economic diversification. The city's economic vitality led it to grow from 1,000 people in 1850 to 45,000 in 1880. Its bourgeoisie grew more cohesive. By the 1880s, it was tied together by institutions such as the board of trade, memberships in private clubs, shared directorships on corporate boards, and, of course, intermarriage. By 1880, there were forty men who dominated the Scranton regional economy. They were divided among original settlers, pre-1850 industrialists, post-1850 industrialists, and men in commerce and the professions. Scranton's economic bourgeoisie in 1880 did not have the longevity and continuity enjoyed by the Wilkes-Barre bourgeoisie—only a small portion of the economic leaders in 1920 had roots in the generation of industrialists from the preceding era (ibid., 79, 43ff).

Like the Wilkes-Barre bourgeoisie, Scranton's upper class dominated not only the city's economy but also the economy of a resource-rich hinterland. And, as in Wilkes-Barre, Scranton's bourgeoisie held joint investments in coal and railroads with extra-local capitalists as well as separate investments in local industries both independent of and tied to coal mining.

Mauch Chunk, while not a mining town itself, was the center of the anthracite coal trade in Carbon County, which, itself, was part of Pennsylvania's Lehigh Valley. Mauch Chunk was one of the United States' earliest company towns. From its founding in 1818 until 1831, the town was the property of the Lehigh Coal and Navigation Company, which was the sole proprietor of all building and properties there. Residents and businesses leased from the company. The company was a paternalistic enterprise, supplying residents with a wide variety of services and goods. In 1831, the company began selling property in the town.

Anthracite coal had been discovered by Moravian settlers along the Lehigh River as early as 1746, near what would become Summit Hill. It was used in forges in blacksmith shops and shipped to the Moravian town of Nazareth in Northampton County (Campion 1997, 79).

As demand for anthracite increased, three entrepreneurs in the Lehigh Valley, Josiah White, Erskine Hazard, and George Hauto, sought to meet it. At first they explored the possibility of gaining access to coal deposits in Schuylkill County, but they were frustrated in their attempt by the existence of the Schuylkill Navigation Company (ibid., 80). Therefore, in 1818, they turned to exploiting the deposits of anthracite along the upper Lehigh River. They received navigational rights along the Lehigh, giving them total control of shipments of coal on that stream. By 1820, the first loads of coal were being shipped on the Lehigh to Philadelphia. In 1821, the entrepreneurs merged their two companies (the Lehigh Coal Company and the Lehigh Navigation Company) into one enterprise, the Lehigh Coal and Navigation Company.

By 1829, the new company had built a canal from Mauch Chunk to Easton, Pennsylvania, where it was connected to two other canals, one that went to Philadelphia and the other that crossed New Jersey and went on to New York City. In 1855, the Lehigh Valley Railroad reached Mauch Chunk, opening new transportation opportunities for Carbon County coal and consolidating Mauch Chunck's position as an emerging dominant player in the trade. There followed a rapid growth in railroads, linking the Lehigh Valley to Wilkes-Barre, Hazleton, Pottsville, and other coal centers. Mauch Chunk came to be linked to the New York City region by the Central Railroad of New Jersey (the Jersey Central, as it was called in the coalfields). There also were rail connections to Boston and other points in New England and upstate New York.

Coal and its transport made Mauch Chunk the industrial center of the Lehigh Valley, with levels of wealth exceeding that of the much larger cities of Allentown, Bethlehem, and Easton, all of which were outside of the coal belt (Folsom 1981, 120–21). Mauch Chunk and Carbon County had a bifurcated bourgeoisie. One faction was organized around the Lehigh Coal and Navigation Company and the other around Asa Packer and the Lehigh Valley Railroad. These two groups came to control the development of Mauch Chunk and Carbon County, but each had its own agenda, which was not always complementary to the other. While they were in Mauch Chunk, the two groups not only dominated the local economy, including banking, but the political and social systems as well.

The Packer group eventually shifted their investments and interests to South Bethlehem, where they created a network of institutions to serve their members. Like the Scranton bourgeoisie, the Packer circle had trouble maintaining its class position past the death of the founding generation. By 1893, the railroad and the iron works, the core industries of the Packer group were in serious trouble, undermining the solidarity of the group in South Bethlehem (ibid.).

In sum, by the end of the nineteenth century, the two largest cities in the coal region, Scranton and Wilkes-Barre, had clearly defined and self-perpetuating bourgeoisies. These bourgeoisies exercised political, social, and economic hegemony over their respective urban economies and the two cities' extensive hinterlands of smaller cities and large agricultural areas. The presence of these bourgeois classes, which were linked in terms of their overall interest in promoting the coal industry and regional development, despite their competition for regional dominance, was both a product of the nature of urban and capital investment and development and a cause of such development and investment. They made sure that their respective sub-regions were maintained as attractive investment arenas for extra-local capital compared to the coal regions that lay to the south of Luzerne and Lackawanna Counties. The bourgeoisie in Scranton and Wilkes-Barre facilitated the development of mining as a corporate activity. Working in concert with capitalists primarily in New York City, they organized coal production in such a way that after the Civil War, the dominant center of output was in the two northern fields. In the southern and central fields, the failure of a coherent bourgeoisie to form and take control, either on its own or in conjunction with extra-local capital, led to a more disorganized system of production and class political control over the regional economies.

Chapter 3

Worker Rebellion and the Problems of Worker Solidarity

From the earliest days of the coal industry, the workers were a militant lot. The first recorded strike in the industry over wages took place among canal boatmen in 1835. During this period, the development and operation of canals was essential to the maintenance and expansion of the coal trade. The strike was accompanied by riots as strikers tried to prevent non-strikers from working. Civil authorities did nothing to stop the strike and end the accompanying violence until a group of prominent Pottsville citizens tried to send a boatload of coal through the blockades the boatmen had established. Several of these citizens were severely assaulted, at which point the sheriff intervened and arrested seventeen of the strikers. This was followed by a march on the county seat of Pottsville by about 300 people to free the seventeen men arrested. The sheriff arrested the leaders of the march and drove off the remainder of the strikers, bringing to an end a three-week shutdown of the coal trade (Luther 1881, 49).

Miners, like the boatmen, used the strike as a weapon to defend their wage levels and improve their working conditions. The wage rate in mining was highly variable across jobs, across production points, and across time. In boom times wages were higher, but never enough to support a family. In times of economic crisis wages would plummet and unemployment would soar, threatening families with destitution. During the nineteenth century, strikes were not a very effective weapon against employers due to the fact that workers were divided and it was hard to get them to act in solidarity due to repression, both public and private.

While workers stood in a conflicting relationship with a large number of employers, conflict was especially intense between workers and the Philadelphia and Reading Railroad and its mining subsidiary, the Philadelphia and Reading Coal and Iron Company. The company ran its first train from

Philadelphia to Mount Carbon on January 1, 1842, and the railroad officially opened for business on January 11, 1842. An immediate effect of the opening of the railroad was a drop in transportation costs for coal. This reduction brought no benefits to producers who, in their competitive struggles against one another, lowered their prices. Competition and a reduction in demand kept the petty producers in the Northumberland and Schuylkill County fields from raising their prices for tonnage. The price reductions by the producers and drops in freight rates put coal prices near an all-time low in the industry's brief history. Producers were continually at the mercy of the haulers in setting their prices. In some parts of the industry, the dependency of producers on the railroads was formal, with the transport companies owning the mines. In other places it was informal, with transporters holding monopolies over given areas that gave them the power to set coal prices and define production amounts.

In 1842 in Schuylkill County, the production side of the industry had its first strike when about 1,500 men left their jobs over wage cuts (ibid.). They soon were joined by others, and riots ensued as the strikers tried to block other men from going to work. Local authorities suppressed the strike in about three weeks, and all of the men, except for those dismissed because of their leadership roles in the strike, returned to work. This strike, like that of the boatmen in 1835, was an example of pre-political collective action and was quickly defeated by the coal operators, in part because significant portions of the workforce did not want to strike (Keil 1988). The strike, however, proved to be the opening salvo in labor battles that lasted throughout the nineteenth century and well into the twentieth in the mining industry, as well as in other regional industries.

The strike was a product of a drop in miners' wages in 1842 as a result of falling prices and slack demand. Miners' wages were lowered to $5.25 per week. For the unskilled laborers, who assisted miners and the mining process, wages fell to $4.20 a week. At many of the mines, the workers were paid in kind (i.e., in coal or in products at the company stores); and as a result, "There was scarcely cash enough paid at the mines to bury the dead" (Luther 1881).

Strikers invaded a number of cities and towns and brought work to a halt. The voluntary militia was called up to restore order and end the riots at several of the collieries after the sheriff's forces proved incapable of handling the situation. The strike lasted nearly a month with the men gradually returning to work on company terms. This was clearly an example of another pre-political action (Hobsbawm 1959) in that the workers had no clear agenda other than to relieve their economic pressures. They did not have a political objective such as over-turning the prevailing system of production and restructuring of the relations between capital and labor. Thus, it was similar to a classic peasant rebellion and had the same effect on the middle classes of Schuylkill County that such rebellions had on European landowners and petty entrepreneurs—a demand for the restoration of order at almost any cost, which the state was willing to do. The appearance of miners in their work regalia carrying the tools of their trade in cities such as Pottsville terrified the urban middle classes and the petty producers who lived outside of the world of mining patches and coal collieries. They had never seen anything like this, and were terrified and incensed by the spectacle of striking miners invading their space.

Things were not much better in the coal trade in the southern fields in 1843. Competitive pressures still worked their will on the petty producers in the southern anthracite fields, driving many of them out of business. Again, workers bore a greater share of the economic burden of 1843 than did operators. The 1842 strike had done little to improve their position, given that the producers had held out until state repression and hunger drove the miners to surrender to the companies' wage policies.

In 1844, workers on the Schuylkill Valley Railroad struck in the middle of December. This strike, like that of the miners in 1842, turned violent, and four companies of the voluntary militia were called out to suppress it (ibid.). In 1847, workers on the Panther Creek Railroad in Carbon County, which had been carved out of Northampton and Monroe Counties in 1843, struck for higher wages. The railroad's owners paid strikers the wages they were owed for past work and dismissed all of them (Campion 1997).

From 1844 to 1846, prosperity returned to the coalfields as demand for coal rose. As demand rose, wages increased, employment dropped, and a large surplus of coal was generated. With rising demand and increased supply, which was greater than the demand, the consequent oversupply led to wage cuts and increasing unemployment. This was the result of the structure of competition between petty producers and between the Schuylkill Navigation Company and the Reading Railroad for access to coal to haul to market.

In 1847, the Schuylkill Navigation Company and the Reading Railroad tried to reach an accommodation that would reduce the disastrous competition between them by working out a shared allocation of coal to be transported. The Reading offered to permit the Schuylkill Navigation Company 400,000 tons of coal to haul, which was far below the Schuylkill Navigation Company's hauling capacity, while the Schuylkill Navigation Company demanded 500,000 tons instead. Negotiations broke down. In 1849, the Schuylkill Navigation Company and the Reading Railroad agreed to divide shipping, with the former granted the right to haul 600,000 tons of anthracite, though the actual tonnage hauled was only 489,208 tons that year.

Coal prices still remained depressed, due to a continuing oversupply of coal and its consequent fall in prices. The larger operators, who were responsible for three-fourths of the production in 1848, decided to suspend production from March 19 to April 7 in order to reduce coal supplies (Luther 1881, 56). The suspension eventually lasted until May 2, a total of seven weeks. The suspension put extreme economic pressure on mineworkers and their families.

On the day work was supposed to resume, the operators were met by a strike of miners and laborers for an increase in wages. This strike was led by the Bates' Union, the first union in the anthracite coal industry. The unionized workers engaged in violence against those men refusing to honor their picket lines. The strike lasted until May 21, when it collapsed under pressure from the operators (ibid., 57). With the end of the suspension and strike, production boomed. Operators voluntarily suspended production for two more weeks on June 23 in order to control the oversupply of anthracite coal in the hope of raising prices.

Little is known about the Bates' Union other than that it was limited to a small number of production points, appears to have been mainly a Welsh mineworkers' organization, and died off in 1850. Its significance lies in the fact that the organization was the first union in the anthracite mining industry, uniting mineworkers of all occupations and skill levels into one union (Greene 1968; Keil 1982, 1988).

Between 1850 and 1868, mineworkers continued their efforts to build stable unions capable of delivering higher wages, more stable employment, and safer working conditions. None of the unions lasted for more than a few months, and none appears to have spread beyond a few of the pits in either the northern or southern fields.

By 1850, it was an accepted fact in the coal trade that one of the major threats to profitability in the Schuylkill region was the large numbers of petty producers. The small producers "were in a great measure the cause of the ruinous prices that so frequently prevailed." The president of the Reading Railroad was said to have commented that the mining of coal would never become profitable until the small operators were driven out of business (Luther 1881, 58).

The presence, as of 1850, of over one hundred producers in the Schuylkill field resulted in high levels of output. Given their lack of organization, they failed to effectively control the supply and marketing of Schuylkill coal, thus minimizing profitability. Producers in Schuylkill seemed incapable of organizing production on a regular basis and joining together to regulate the trade during times of crisis, when profits were squeezed or wiped out by oversupply and competitive marketing.

Profits in the southern anthracite region were further squeezed in 1851 when the larger corporate producers in the northern region banded together to compete against the fragmented producers in the southern anthracite fields.

Schuylkill producers tried to unite and negotiate with the northern cartel to achieve an equitable distribution of production and coal marketing, but they were not successful in establishing a stable arrangement for dividing the amount of coal to be produced and marketed across the anthracite regions.

In 1852, coal demand rose precipitously, clearing the market of the surpluses that had existed in 1851. Output, prices, and wages also rose significantly in early 1852, only to fall in the autumn of that year, as demand dropped dramatically. Drops in production continued into 1853; as a result downward pressure on wages occurred. Miners and mine laborers responded with a number of strikes and demonstrations. These strikes lacked the organization that a union would have brought to them. In a sense, they merely were spontaneous outbreaks of labor protest over wage reductions. They were more examples of pre-political (Hobsbawm 1959) action taken to relieve immediate economic pressures, with no long-range plan to restructure the industry and the labor processes in it.

By 1854, the coal trade had returned to prosperity across all of the anthracite fields. Prices rose for operators and shippers, and miners' wages rose significantly. In 1856, another attempt was made in the Schuylkill field to rationalize the coal trade, when a number of the larger producers joined together to form a cartel. They hired a man named Samuel Sillyman to serve as president of their "coal association," which held weekly meetings to monitor and make adjustments in the coal trade and deal with continuing increases in competition from producers in the Luzerne and Lackawanna fields (ibid., 61).

In 1857, the operators in the Schuylkill field hired John Tucker, who had just resigned as president of the Reading Railroad, to take over management of their trade association, in the hope that he could impose discipline on the producers. Tucker was not successful in his efforts to bring discipline to production and the marketing of Schuylkill County coal. Matters were made worse with the financial crash at the end of 1857, which produced a collapse in the southern field's coal trade association and left each operator to fend for himself in the competitive market for coal.

In the meantime, competitive pressures from the Luzerne and Lackawanna fields and from Carbon County intensified, resulting in further economic anarchy in the coal trade. Individual operators and transporters in the southern fields could not compete effectively against the giant integrated mining and transportation companies that were coming to dominate in the other fields, especially in Luzerne and Lackawanna Counties. For the first time in the history of the Pennsylvania coal trade, in 1861 the Schuylkill region was surpassed in production by the northern region. This greater production from the Wyoming and Lackawanna fields lasted almost to the demise of the coal industry.

The Civil War years (1861–1865) constituted a boom time for coal operators and transportation companies in all of the anthracite fields, as the government and industries involved in the war effort bought ever-larger amounts of anthracite coal. As production ratcheted upward, so too did miners' wage demands. In 1862, there were frequent strikes and acts of violence perpetrated against the coal producers and the workers who remained on the job during walkouts. According to P.D. Luther, there was a great deal of class envy on the part of the middle classes toward the miners and the wages they were earning. The skilled workers outside of the industry, petty producers, white-collar workers, and lower-middle-class entrepreneurs could not understand how such lowly beings as miners could be making more money than them (1881, 64).

The 1860s were marked by violence in the coalfields. Some mineworkers used intimidation tactics against their fellow workers and the companies in order to drive up wages, which further incensed the region's middle classes (ibid.). Conditions became so bad that public authorities and mining companies brought in a part of the New Jersey 10th Regiment in 1864 to Cass Township to assert control over the region. Some of the violence in the coalfields at this time was related to Irish anti-draft riots and the fact that rich men could buy themselves out of military service, while poorer people, such as miners and their laborers, were compelled to serve in the Union's military forces. Other violence was grounded in the relations of production and in interethnic competition.

As its profits increased dramatically during the Civil War, the Reading Railroad set about leasing and purchasing smaller railroads to increase its control over hauling operations in the Schuylkill region and thereby expand its indirect control overproduction. It also set about building new lines. It relied heavily on British capital to finance its expansion, going deeply into debt as a result of its borrowing practices. As time went on, the railroad had increasing trouble servicing its debt. It eventually went bankrupt and fell into receivership in 1880.

The year 1864 saw a consolidation in the production of coal in the Schuylkill field. With prices for coal lands rising rapidly, a large number of petty producers decided to get out of the business with huge profits. Consolidation, however, did not completely eliminate petty production in the southern anthracite fields. A number of petty producers continued to operate and new ones entered the region to take advantage of the large profit during the Civil War, which prevented the complete rationalization of production under the hegemony of a cartel of larger operators.

In early July 1864, engineers and brakemen on lateral railroads that fed the main lines, such as the Reading, went on strike in the Schuylkill region demanding wage increases. Given the negative impact that this walkout had on the war effort, the U.S. government seized control of the Reading and its lateral lines and replaced the striking workers. The workers surrendered after two weeks and returned to work without realizing any of their demands.

With the end of the Civil War in 1865, many industries that heavily used anthracite coal cut back on their own production, and demand for anthracite collapsed. Prices fell below the cost of production for anthracite. Operators demanded wage cuts of 25 to 33 percent from the mineworkers (ibid., 66). The miners resisted by striking. The strikes in Schuylkill lasted for about two months, when the miners finally surrendered to the operators' demands. In the Lackawanna and Wyoming regions to the north, the strikes lasted about ten weeks before the miners caved in and returned to work in defeat.

In 1866, demand for anthracite coal fell again. Once again, the operators turned to the miners and laborers to reduce wages. The workers again resisted and, this time, were partially successful in fighting wage cuts. Beginning in July 1868, the workers struck through September 1 for the eight-hour day, with pay for ten hours. This strike, like others that preceded it, was characterized by riots and violence against company property and against workers who failed to comply with the demand for a walkout. The companies refused to concede an eight-hour workday, which had become a demand of large numbers of workers across the country, and the workers returned to the mines in defeat. Ironically, this strike saved the coal trade for the year by decreasing supply and increasing prices for the backlog of coal that had been produced. As the market for anthracite cleared, and with demand for and the price of coal increasing, workers demanded and received higher wages.

Mineworkers established the WBA in 1868. It carried out a successful strike in 1869 and was instrumental in initiating mine safety legislation (Keil 1988; Miller and Sharpless 1985). Mine safety had become an especially critical issue following the Avondale Disaster in 1869 just south of Plymouth, Pennsylvania, in which 108 men and boys were killed in a mine fire. There had only been one source of egress from the mine and one air hole, making the mine a death trap for those working in it. The Nanticoke Coal and Iron Company, a subsidiary of the Lackawanna Railroad, operated that mine.

The WBA was the first real union in the anthracite coalfields. Founded on July 23, 1868, it stands out in American labor history for two reasons. It was the first union to unite workers from all of the anthracite production fields, and it was the first instance of a major American union organized along industrial rather than trade or craft lines. It was formed as a direct result of a strike at Girardsville in Schuylkill County. A year later, the WBA carried out its first successful general strike in the industry and it subsequently conducted strikes in 1870 and 1871.

Along with carrying out labor actions, the WBA embarked on an ambitious legislative campaign and political agenda, and succeeded in having the state legislature pass a number of mine safety bills. Mining was a dangerous job well into the twentieth century. There were scores, if not hundreds, of deaths each year from preventable accidents. In some of these, whole families of men and boys would be wiped out. Those who died would be brought home in mule wagons, and their bodies would be deposited on porches, with no formal expression of sympathy for the death and no compensation for the survivors. Over 30,000 lives were lost directly over the course of the anthracite industry's history. This does not take into account those whose lives were cut short by "black lung," a form of emphysema brought on by constant breathing in of anthracite dust.

The year 1869 saw a number of significant developments for the Schuylkill County coal business. Once again there was a major strike on the part of miners and mine laborers, including all workers in the industry, except for those in the Lackawanna field.

In April 1869, Franklin B. Gowen was elected president of the Reading Railroad. Gowen's appointment coincided, a few months later, with the attempt to form another board of trade in the Schuylkill anthracite fields. In addition to regulating production, this board took upon itself the task of negotiating with the mineworkers over wages and working conditions in the hope of creating a common negotiating strategy and policy position among the owners with respect to mineworkers' wage and other demands dealing with working conditions.

Mining operations were struck April 1, 1870, through August 1, 1870, over a wage dispute between operators and miners. The miners wanted to maintain the rates of 1869 ($3.00 per ton), while the operators wanted to cut this rate, arguing that Schuylkill coal was not competitive at that price. The board of trade and union agreed that Gowen would arbitrate the dispute. He proposed maintaining the $3.00 per ton wage rate, but pegged changes in the rate to a sliding scale of the price of coal sales at Port Carbon. Both the union and the board of trade agreed to what came to be called the "Gowen Compromise" (Luther 1881, 68). Under the terms of this agreement, the miners ended up taking a wage cut as coal prices at Port Carbon fell with declining demand for anthracite.

On November 7, 1870, the WBA's negotiating committee met with representatives of the Schuylkill anthracite board of trade to draw up wages for 1871. An agreement was reached that set a base wage rate keyed the price of $2.50 per ton of coal at Port Carbon.

Shortly after becoming president of the Reading Railroad, Gowen, the scion of a prosperous Irish Protestant liquor wholesaler from Philadelphia, began to take steps to convert the railroad from exclusively hauling coal into a substantial and significant coal producer, in direct violation of the company's initial charter. At first, Gowen entered into the production business by leasing lands, but he soon discovered that the real money was to be made in owning mines outright. But, the railroads' corporate charter prohibited this. That soon would end.

The WBA directed another strike in 1871 that largely failed due in part to the inability of the union to hold the workforce together. One of the main obstacles that the mineworkers faced in constructing durable and effective unions was the divisions and rivalry among various nationality groups for jobs in the mines. The different ethnic groups working in the minefields not only had to overcome interethnic rivalries, but also intra-communal cleavages and hostilities as well. Few of the ethnic groups that arrived in the United States in general and the coalfields in particular had a clear collective identity. For example, German identity was organized more along the lines of religion and feudal region than from the "fact" of being a "German." Similarly the Irish, pegged their identity to one of Ireland's six historical regions and within, regions, to a particular county. A man from Cork was a man from Cork more than he was a man from Ireland. This fracturing of identities led to internecine conflicts within immigrant populations, as well as between them.

In most cases, the structural solidarities of ethnicity and ethnic consciousness emerged after the fact of immigration and were not brought with the immigrant to America (Yancey et al. 1976; Oppenheimer 1974; Reich 1977; Cox 1948). Ethnic identity was as much imposed from the "outside" as it was constructed from the "inside." Once constructed, however, it was one source of cultural succor and economic support in the new world the immigrants encountered.

The ethnic communities, however, were important not only for the personal assistance they provided to their members. They also were fashioned into collective weapons that could be harnessed to the mineworkers' cause and that often figured prominently in mineworkers' struggles against their employers (Greene 1968; Keil 1982, 1988; Miller and Sharpless 1985). In these struggles, ethnic communities frequently blended non-or pre-industrial practices, such as terrorist actions, with more modern strategies and tactics, such as the strike and direct political action. The workers' movement adopted forms of protest that were characteristic of ethnic communal conflicts and synthesized them with more modern forms of protest, collective action, and class struggle and conflict, and adopted conceptions of economic and social justice that were heavily influenced by the moral economy of rural, peasant societies (Keil 1988, 26–27).

It was not only European immigrants who created identity-based communities. Associations were formed throughout the coalfields to protect and advance the interests of native white workers, the majority of whom were Protestants. Among the more important of these various lodges were the

Masons, the Odd Fellows, the United Order of American Mechanics, the Junior United Order of American Mechanics, and the Knights of St. Pithias.

For the Irish, the Ancient Order of Hibernians was a critical institution. Economic and political competition often led to violent conflicts such as those among Hibernians and members of organizations such as the Masons and the Odd Fellows, as well as Welsh and German mineworkers. The majority of these latter organizations were made up of Protestants, and they contained not a small number of mine supervisors and superintendents. The Hibernians were exclusively Irish and overwhelmingly Catholic, and its membership was heavily made up of miners and mine laborers, as well as non-mineworkers and petty bourgeois entrepreneurs.

The Molly Maguires, a secret society that was said to exist within the Hibernian order, were reputed to use extreme violence against non-Irish workers and supervisors in the mines. They operated in the coalfields from the 1850s through the late 1870s (Bimba 1932; Coleman 1936; Lewis 1964; Aurand 1971; Keil 1982, 1988). They directed their attacks at fellow Irishmen who refused to show solidarity with their political and economic causes and at American, Welsh, and German mineworkers who they believed were collaborating with mine operators. They also challenged operators and supervisors who the Mollies felt had singled out the Irish for hyper-exploitation and mistreated Irish workers. The Mollies reputedly were far more active in the central and southern coalfields of Carbon, Northumberland, and Schuylkill Counties than in the Luzerne and Lackawanna fields. The Irish were not alone in developing secret "terrorist" organizations—the Welsh, the Cornish, and the Germans, among others, also formed such groups and used them as defensive and offensive organizations, especially against the Irish (Keil 1988).

The Molly Maguire phenomenon was a direct carryover from forms of Irish peasant resistance to Irish and Anglo-Irish landowners and the so-called rack men who collected rent from peasant tenant farmers and cottagers in Irish villages. In Ireland, the phenomenon of secret terrorist societies was referred to as "Ribbonmen." The Mollies were first mentioned in 1857 as operating in the coalfields in a story printed in the anti-labor newspaper *The Pottsville Miner's Journal.* The paper's publisher was Benjamin Bannan, a major activist in the newly formed Republican Party. In the elections of 1856, Bannan unsuccessfully had tried to break the ties between the Irish and the local Democratic Party. Unable to do so, he reprinted an article from a Philadelphia newspaper that accused an Irish organization, the Molly Maguires, of having engaged in massive voter fraud on behalf of the Democrats. The Mollies were linked to crimes in Schuylkill, Northumberland, and Carbon Counties from time to time after Bannan's article, but they really came under severe attack in the 1870s by the Reading Railroad.

In 1871, at Gowen's behest, a state legislator introduced a bill that would have allowed the formation of a company called the Franklin Coal Company. This new entity would have had the right to own and operate mines. The bill provided that any railroad company could take control of a mining firm by either buying its stock or bonds. Opposition rose in the legislature to this bill, and it did not pass. Opponents feared that if enacted, the bill would drive small producers out of business and create a monopoly over coal production and distribution in Schuylkill County.

Eventually, however, Gowen and his legislative allies did manage to obtain passage of a similar bill that allowed for the formation of the Laurel Run Improvement Company, which soon fell under the control of the Reading Railroad, putting the railroad directly into the coal production business for the first time. By the end of 1871, the railroad owned over 65,000 acres of coal land, one of the largest coal holdings in the entire industry.

In legislative hearings in 1872, called over Gowen's raising of coal hauling prices, he put part of the blame for price increases on the depredations of Molly Maguires, which he linked to the WBA. This was the first time Gowen had alluded to the presence of the Mollies in the Schuylkill coalfields, and the first time they were linked to a trade union.

In 1872, shortly following his price increase, Gowen took steps to unite other major independent producers across production regions into a production/distribution cartel. He held a meeting with Asa Packer, president of the Lehigh Valley Railroad, and Charles Parrish of the Lehigh and Wilkes-Barre Coal Company. Gowen, Parrish, and Packer agreed not to go after each other's holdings and markets, deeming such competition to be highly destructive to their individual interests. At the same time, however, they agreed that competition among the three to acquire the property of smaller producers and railroads was permissible.

Eighteen seventy-four was a hard year for mineworkers. The Anthracite Trade Association, an organization made up of the larger anthracite producers formed for the purpose of regulating the trade and setting wage rates in the industry, met in August 1874 and announced that there were going to be substantial across-the-board wage cuts in 1875. This would come on top of already deplorable conditions in the mining industry. Mayor Mathew Loftus of Scranton wrote a letter to the *New York Herald* reporting on the horrendous conditions in the northern mine fields. The mines had been shut all summer, and by December, the mayor reported, they were working only half to two-thirds time. Thousands of people were described as destitute. The *Luzerne Union* reprinted Loftus's letter for local readers and also printed a letter that the Roman Catholic bishop had written to the *Herald* about generally poor conditions in the minefields. The *Luzerne Union* likewise published a letter written by a miner named O'Halloran, the president of the regionally based

National Miners' Association of Luzerne County, that accused the companies of blacklisting union members and dissident workers, price gouging at company stores, charging fees for unneeded or unused medical services, and raising prices for goods and services at company stores in the mining patches at the same time they were cutting wages. O'Halloran went on to claim that mass starvation was a real danger. Nothing much came of these complaints. There does not appear to have been a groundswell of support for the miners and their families in the general community or in the business communities of Luzerne County.

By 1875, the WBA had changed its name to the Miners' and Laborers' Benevolent Association (MLBA), specifically identifying itself as an industrial union rather than a trade union. The year proved to be disastrous for the MLBA and its members. Throughout the coalfields, wages were cut by 20 percent. In response, the union called a strike, which lasted for six months. During that time, the miners exhausted their personal resources. They consumed their small gardens' entire produce and killed-off small animals, such as chickens, ducks, geese, rabbits, pigs, and so on. The mineworkers and their families were reduced to foraging and hunting to supply themselves with food. After six months on strike, the miners returned to work, beaten down by the companies, and accepted the wage cuts. The miners and their families were literally starving and suffering from severe malnutrition.

In addition, mineworkers faced massive repression by the Pinkertons, a private detective agency frequently employed by the Reading to disorganize the workforce, and the Coal and Iron Police, a private police force controlled by the Reading. The latter attacked strikers and terrorized entire mining communities, with little or no response from state authorities. The miners, however, organized their own resistance and fought back against the Coal and Iron Police, launching attacks against coal and railroad properties. Still, the operators ultimately defeated the mineworkers (O'Bannon et al. 1997, 66). With the collapse of the strike of 1875, the MLBA disappeared from the coalfields as an organized force.

Unknown to the mineworkers when they went on strike, the Reading and other key producers had stockpiled coal in Philadelphia and other cities in preparation for a long work stoppage. There was enough coal held in reserve that the producers could wait-out the strikers and starve them into submission, which they did. At its peak, the MLBA represented about 80 percent of the mine workforce, with the remainder of mineworkers represented by smaller unions or unrepresented (ibid., 63).

Despite its terrible defeat in the strike of 1875, the MLBA taught the workers a number of things. The most important lesson it offered the miners and mine laborers was that their economic interests were more effective as a basis for organizing than their interests grounded in ethnicity, religion, skill level, mining field, and so on. It also showed them the advantages of organized, political and economic action rather than pre-political protest.

When the MLBA collapsed, mineworkers reverted to pre-political, particularistic social networks, associations, and institutions in order to protect themselves. Ethnic and political antagonisms within the mine workforce reemerged, compounding workers' political difficulties. Without the MLBA to provide a platform for collective action, the mineworkers found it next to impossible to organize on a global basis across all of the minefields. The mineworkers had seen how the lack of effective organization, even in the presence of the WBA, had led to major defeats in the strikes of 1870 and 1871.

Gowen and his partners in the Schuylkill cartel joined with the major corporate producers from the northern fields and established an agreement that allotted quotas to the various producers. This cartel, however, collapsed in 1876 under the pressure of oversupply and the consequent rapid fall in coal prices.

With his major competitors subordinated or allied with him and the MLBA defeated, Gowen felt it was time to go after the Mollies. He hired the Pinkerton Detective Agency to penetrate the Mollies and report back to him on their plans and activities. The Reading also tried to crush the Mollies using the power of the notorious Coal and Iron Police, who had statewide law enforcement powers, even before the creation of the Pennsylvania State Police in 1905.

In a series of trials in the late 1870s, at which Gowen served as a special prosecutor, several Mollies were tried and convicted of murder and hanged for alleged criminal offenses. Among those hanged was Black Jack Kehoe, a prominent Democratic politician and former mineworker in Schuylkill County. Kehoe had been an ardent opponent of the draft during the Civil War and reputedly had led several riots when federal and state authorities tried to enforce the draft against poor, primarily Irish mineworkers, while permitting the children of the coal aristocracy and other segments of the bourgeoisie and petty bourgeoisie to buy their way out of military service. Kehoe (although he subsequently was pardoned by Governor Milton Shapp in 1979) was alleged to have led a group of Mollie nightriders that specially targeted property of the Reading Railroad.

The Molly Maguires' trials were flawed. In many cases, jurors spoke only German and had little, if any, understanding of English. In addition, many juries were composed of members of the Odd Fellows, an organization whose members were often targets of the Mollies. Moreover, lawyers from the major coal-producing companies frequently led or assisted in the prosecution. The Pinkertons and the Coal and Iron Police, hardly unbiased sources, led the investigations. Finally, the judges at the various trials had close personal economic and political ties to the railroads and coal companies (Crown and Major 2003, 207, 208).

The supposed crushing of the Mollies did little to stop labor unrest in the Pennsylvania coalfields. Mine work, except for a handful of so-called "contract miners," was always a marginal economic activity for the miners. The coal companies assembled an enormous workforce as a means of generating downward pressure on wages. This "reserve army" was composed of a heterogeneous lot. During the Civil War, it was composed of Irish, Welsh, Scots, English, Germans, and Americans; after 1880, it also would include groups from eastern, southern, and central Europe, especially from Italy and Poland. These new streams of migration complicated the miners' attempts to achieve a viable form of organizational solidarity.

Just as the mineworkers struggled to overcome the disorganizing effects of competition among themselves for jobs and wages, the companies tried to find ways to lessen the impact of intra-industry competition and regulate the supplies of coal being brought to market (Healey 2007). A coal combination was put into place on January 22, 1873, and lasted until August 21, 1876. This was a formal agreement among the producers to allocate coal allotments. Following the demise of this combination, a period of free competition lasted until 1878, when an agreement was reached that covered that year. In 1879, there was no restraint on competition. Between 1880 and 1884, another informal agreement prevailed among the coal companies. On December 31, 1884, the corporations established terms to control overstocking of the market. This policy lasted through March 1887, though it almost ended in 1886. The influence of J. P. Morgan created the conditions for it to last through March 1887. The repeated failures of these agreements attest to the brutally competitive nature of the anthracite industry and the inability of the largest producers to maintain the discipline necessary to moderate it.

The passage of the Interstate Commerce Commission (ICC) Act seemingly outlawed the formation of coal combinations after January 1, 1887. As a result, the coal companies turned to more informal agreements that were constructed and maintained by their principal sales agents. These arrangements did not have much staying power. For example, in 1892 the Reading leased the Lehigh Valley Railroad and the Central Railroad of New Jersey, which gave it control over 70 percent of the anthracite trade. These arrangements fell apart after the Reading went into bankruptcy for the third time in twelve years. The haulers retuned to informal agreements in 1895, but this also turned out to be unsuccessful, and the railroads formulated an unsigned formal agreement on February 1, 1896, in order to avoid being detected by the federal government under the provisions of the ICC. The agreement was not renewed for 1897, and the companies engaged in unrestricted production, which lasted until the middle of 1898. After 1898, Morgan and other major investors began to build "a community of interest" bound together by cross-ownership of stocks and interlocking directorates (ibid., 252). After the 1880s, the coal region became flooded with workers drawn from eastern and southern Europe: Poles, Russians, Ukrainians, Serbs, Croats, Slovaks, Slovenians, Slavonians, Czechs, Ukrainians, Hungarians, Italians, Sicilians, with a scattering of Greeks, Christian Arabs from the Levant, and Jews from Germany, Russia, and eastern and central Europe. The latter three groups tended not to be involved in mining.

The arrival of these new groups exacerbated interethnic tensions in the region, especially insofar as the groups already working in the mines tried to restrict newcomers to unskilled positions in the mines and ancillary industries. These newer groups were also believed by populations already in place to be people who could not be trusted to act in solidarity with other workers, given their desperate poverty and "strange" cultural practices. Hundreds of families from the new immigrant streams were living in mud huts or caves and secured their food from foraging and game poaching. The men from these groups were relentless in acquiring the skills that would enable them to take the state certification exams that would allow them to be employed as miners, rather than as laborers or in other forms of unskilled work. The indigenous workforce fought every effort of the new immigrants to become licensed miners, including having state representatives pass discriminatory legislation against them (Greene 1968; Keil 1982). These newly emerging interethnic tensions became one more of the many obstacles faced by workers trying to organize in the anthracite industry.

But it was not just ethnic tensions that served as an obstacle to organization. To explain union failures, one must also take into account competition among workers across the various mining fields, divisions within the workforce by skill levels, differing ideologies among the workers, and the active repression of the workers' movements by private and public power, often acting in tandem. This active repression taught the workers the importance of gaining political control over the state apparatus in the coalfields.

Chapter 4

Industrial Unionism Reemerges in the Coalfields

The absence of an effective union did not prevent workers from striking. Violent and often uncoordinated strikes swept the coal regions throughout the 1870s. The most significant of these took place in 1877. The Wyoming and Lehigh fields' workers walked out in support of striking railroad workers in 1877, at the same time that the Reading Railroad's mineworkers continued on the job. In Scranton and Pittston, and a number of other coal sites in northern Luzerne County, there was considerable support for the striking railroad workers. In these areas, there were several acts of civil disobedience and massive public demonstrations of support for the strikers. The demonstrations were multiethnic protests. Segments of the local bourgeoisie hired thugs and gunmen to suppress the protests and other acts of civil disobedience (Brecher 1972). In what came to be known as the Reading Massacre on July 23, 1877, a militia unit fired on an unarmed crowd that was blocking the Reading's trains. Ten people were killed. The U.S. military forced the Reading Railroad to reopen.

Six demonstrators were killed and fifty-four wounded in Scranton when vigilante gunmen fired into a crowd of protestors. The state sent the militia into Wilkes-Barre and Scranton to restore order. As the militia approached the region, they met massive resistance. Their troop trains were attacked by mineworkers as they moved through the deep mountain passes in the coalfields and it took the militia several weeks before they managed to restore order. During this time, workers from a variety of industries invaded the state capital in Harrisburg and held the legislature hostage. In addition, Wyoming Valley miners invaded private coal lands and opened their own mines, expropriating company coal, prefiguring what would be done on a larger scale during the Great Depression of the 1920s and 1930s. Meanwhile, the Schuylkill men kept working, much to the chagrin and disdain of the Wyoming miners.

One of the factors that kept the Schuylkill men working was their greater dependency on coal operators than was the case in the Wyoming and Lackawanna fields. In the Schuylkill and Lehigh fields, a large proportion of mining families lived in company houses that they rented from the operators. Hence, if they went on strike they faced immediate eviction. They also were heavily in debt to company stores, and these debts were called in during times of strikes. In the Luzerne and Lackawanna fields, more miners owned their own homes and dealt with private merchants, who would carry their debt during times of labor troubles or periods of unemployment.

But it was not only inter-field rivalries and divergent actions among mineworkers that were resurrected after the collapse of the MLBA and made collective action difficult. Interethnic rivalries and political-philosophical divisions within the workforce also reemerged. Sometimes all of these factors came to the foreground, and their coincidence seriously weakened the workers' movement.

For example, during the late 1870s, the Knights of Labor began to organize in the Wyoming and Lackawanna fields. The Knights were especially strong around Scranton, Pittston, and Carbondale, cities to the north of Wilkes-Barre. They had less successful efforts in the Schuylkill field, where German, Welsh, and native miners resisted the Knights because many resented their Irish domination and political-economic philosophy, which had semi-socialist elements and which argued for one massive union representing all workers, including the petty bourgeoisie, and abolition of the wage system for all workers and its replacement by a system of cooperatives. They regarded the wage system as tantamount to slavery.

The Knights' political front was the Greenback-Labor Party, which succeeded at one point in electing a mayor of Scranton (Keil 1982, 1988). The Knights and the Greenback-Labor Party endorsed the nationalization of the United States' mines and coal yards. Many of the non-Irish mineworkers in Schuylkill, however, rejected the Knights because of their pan-labor philosophy, preferring an industry-based unionism. Therefore, a large number of Schuylkill's non-Irish miners joined the Amalgamated Association of Miners and Mine Laborers of Pennsylvania.

This introduced dual unionism into the coalfields, further splitting the workforce. The Amalgamated's organizers came from the bituminous fields into the anthracite region in 1883, and by 1885, they had won support among a relatively large proportion of native, English, Welsh, and German miners in the central and southern coalfields.

The Knights and the Amalgamated competed fiercely and sometimes violently for hegemony in the anthracite region. Nonetheless, they managed to cooperate in the strike of 1887–1888, at least in the southern fields. But, even with their combined power, they were no match for the power of the coal cartel, and both suffered massive defeats, which forced them out of the coalfields. Once again, the miners were disorganized and remained so until the 1890s, when organizers from a small bituminous coal union (the UMWA) moved into the region and began organizing the Wyoming field.

What had set the 1887–1888 strike in motion was the call of the Amalgamated for a 15 percent wage hike. Operators rejected the demands, and 20,000 mineworkers went on strike (O'Bannon et al. 1997, 70). Led by the Reading, some operators and the Reading managed to avoid a strike by giving their workers a temporary 8 percent wage increase, but they refused to recognize the union as the collective representative of the strikers. Operators affected by the strike began evicting families from company housing, denying workers credit at company stores, withholding back pay, and firing strike leaders and leading union activists. Local small businesses and several community leaders, resentful of the mine operators' almost complete dominance of the local economy, took the side of the strikers with contributions to miners' relief funds and assistance for local strike committees (Miller and Sharpless 1985, 218–19).

In December 1887, the Reading and its allies among the independent operators in the southern fields withdrew their temporary offer of an 8 percent wage increase, which resulted in a strike at their operations. Knights of Labor members who worked for the railroad refused to move its coal and were fired en masse. Workers in the northern fields, however, refused to join the strike, considerably weakening the strikers' position in the southern fields. As the strike dragged on, workers' resources to sustain it were depleted. Ultimately both the Knights and the Amalgamated surrendered to corporate demands, and both ended up disappearing as effective forces in the coalfields.

In the 1890s, new union organizers came into the anthracite minefields as representatives of the nascent UMWA. The UMWA started an organizing drive in anthracite in 1894 and, after facing three years of numerous difficulties and setbacks, became a notable presence in the industry (Blatz 1994). It had significant success in the Wyoming field where scores, if not hundreds, of men signed up. This figure eventually would reach into the tens of thousands over the next forty to fifty years. While the Schuylkill, initially, was difficult to organize, partly out of fears of the Reading Railroad's reaction, Schuylkill mineworkers eventually supported the UMWA in the same way and to the same degree that workers in the north did.

However, the UMWA had trouble convincing the English-speaking miners that they could trust and rely on the new immigrants from central, eastern, and southern Europe. The English-speaking miners believed these new populations were willing to work for lower wages, as strikebreakers, and as competitors for the better-paying, safer, and more stable mining jobs. On the other hand, the new immigrants saw the existing English-speaking workforce, especially the contract miners, as one more source of exploitation in their new world. Contract miners were "independent" actors who signed agreements with a mine operator to produce a given amount of coal for a specific amount of money. It was up to the contract miner to hire a crew of laborers to assist him. Contract miners were known for extreme exploitation of their crews. The more intense the exploitation the more money the contract miner made.

So intense was the hostility between English-speaking miners and the new arrivals that it helped produce a schism in the Catholic church in the Scranton Diocese. This developed from 1896 to 1907 and continues to the present. During the summer of 1896, the Sacred Hearts Parish of South Scranton was torn apart by internal conflicts that sometimes led to violence within the parish community. The dissident Polish faction protested the parish's governance by its pastor, Father Richard C. Aust, who was of German background. In October 1896, 250 families seceded from the parish and formed St. Stanislaus Polish Roman Catholic Church. They bought property and immediately began to build a church. They were advised by a young Polish priest from Nanticoke, Father Frances Hodur.

When the committee requested diocesan recognition for the new parish, the newly appointed coadjutor of the diocese, Bishop Hoban, took a hard line and refused to recognize the parish, laying the groundwork for a Polish schism that would lead to the formation of the Polish National Catholic Church. This schism was never over questions of theological doctrine, but, rather over the form of the church polity and the structure of the liturgy. It was a question of lay versus clerical control over church property and parish governance, coupled with interethnic rivalries that produced the split.

On March 14, 1897, Father Hodur left Nanticoke to become the pastor of St. Stanislaus. A week later, Bishop Hoban suspended him as a Catholic priest. In September 1898, Father Hodur agreed to submit to Bishop Hoban's authority, provided three conditions were met: retention of the deed to parish buildings by St. Stanislaus, lay control of finances, and committee approval of pastoral appointments. Bishop Hoban rejected all three conditions.

Father Hodur subsequently went to Rome to plead his case. He was unsuccessful, and on October 15, 1898, Bishop Hoban excommunicated him. The division between the parish and the diocese became more intense when, on Christmas Eve 1900, Polish replaced Latin at the church's liturgical services.

In 1904, a synod of Polish parishes was held, and Father Hodur was named bishop elect of the National Church. Three years later, he was consecrated a bishop in Utrecht.

Eventually, the UMWA's cause of bringing new and old immigrants together was helped when a sheriff's posse, led by Luzerne County sheriff James L. Martin, gunned down a large number of workers, most of whom were Slavs and Magyars, while they were on a peaceful demonstration march, carrying American flags and singing patriotic songs. The men were on their way to shut down a mine run by the Pardee family in southern Luzerne County. Nineteen men were killed and thirty-two wounded, in what came to be known as the Lattimer Massacre (Greene 1968; Novak 1978; Keil 1988; Campion 1997).

The massacre came after a period of labor unrest at a mine owned by the Lehigh and Wilkes-Barre Coal Company, which also operated in Carbon and Schuylkill Counties. The actual labor dispute had begun in a company mine in McAdoo, a mining settlement in Schuylkill County (Campion 1997, 120). The company's superintendent, Gomer James, of Welsh extraction, announced new work rules for the mule drivers, which would have them labor more hours without compensation. The mule drivers went on strike. The day following the walkout, James attacked a young striker, John Bodan, with a crowbar. In return, the strikers began beating James and in all likelihood would have killed him were it not for the intervention of a company foreman. Bodan filed charges against James, who was arrested and imprisoned while awaiting trial. While James waited for his day in court, the striking mineworkers tried to shut down mining operations.

The major coal operators decided it was time to put down the strikes and approached Sheriff Martin to do so. He had been elected with the strong financial and political backing of the coal operators in Luzerne County. There was, however, a problem with Martin's involvement; the protests were not in Luzerne County. Martin declared a state of disorder and secured the same from the sheriffs of Carbon and Schuylkill Counties. He assembled a posse in preparation for protesting marchers crossing into Luzerne County. The coal operators also put together an armed force of company police and an estimated 300 Pinkerton detectives to confront the strikers as they approached Hazleton in southern Luzerne County. Confrontation in Hazleton was avoided when the miners agreed to march around the city rather than through it. The violence broke out as the marchers approached Lattimer and were shot down by members of Martin's posse.

Following the massacre, Pennsylvania sent its 3rd National Guard Brigade to restore "order" and get the miners back to work. They encountered something they could not have expected. Led by a woman called "Big Mary" Stepak, who operated a boarding house, about 150 of the mineworkers' wives blocked their husbands and other men from returning to work.

Martin and his deputies were charged with and tried for murder. They were defended by Henry Palmer, a one-time attorney general of the Commonwealth of Pennsylvania. Part of Palmer's "defense" was that the marchers were "socialists, anarchists, and haters of organized government" (Novak 1978, 232). Martin and his deputies were acquitted of all charges, further inflaming the passions of eastern and central European mineworkers and strengthening their resolve to end their exploitation and degradation by the mining companies and local political elites beholden to and/or allied with the region's coal operators.

The Martin trial and acquittal showed the region's citizens that mine-workers were at best second-class human beings or, even worse, were barely humans. They could be shot down and killed with impunity, whenever the authorities deemed that the mineworkers, even in the smallest way, were threatening institutionalized arrangements of the coal-producing sector and the bourgeoisie's and coal operators' power and domination in the coal regions.

The Lattimer Massacre was a catalyst for further organization by the UMWA in the northern anthracite fields, especially among the newer ethnic groups. By 1900, the UMWA was strong enough to carry out a successful six-week strike, which included workers from the new immigrant groups acting in solidarity with miners from the older groups, at the end of which workers gained considerable success in bargaining with the companies. The agreements expired in 1901, a time of serious strains between mine operators and their workers (Roberts 1901; Greene 1968).

When the agreement expired on April 1, 1901, the mine operators refused to meet with the UMWA or to concede its right to negotiate on behalf of the miners and mineworkers. From January onward, 1901 was marked by labor turmoil. There were reportedly over one hundred labor actions initiated by workers that year (Greene 1968). In 1902, the UMWA presented a set of demands to mine operators. These demands included, among other things, an eight-hour day, a 20 percent wage increase, a significant adjustment in the dockage for waste materials included in the mine wagons hauling coal to the surface, improved working conditions, and recognition of the union as the sole bargaining agent for the mineworkers. The union wanted its demands presented to an impartial board of arbitration.

J. P. Morgan, who had begun to invest heavily in the anthracite coal trade, refused to negotiate on these points and "encouraged" other large corporate operators to follow his lead. Such was Morgan's power in the industry that the other operators were compelled to follow his dictate. On May 12, 1902, the UMWA called a strike, which lasted until October 20, 1902, a total of 165 days. Over 147,000 workers walked off their jobs, effectively shutting down the industry in the anthracite regions. The strike was suspended only after the federal government stepped in and persuaded the operators to accept binding arbitration.

Until the strike suspension, the mineworkers endured horrible economic and social hardships. Many families were on the brink of starvation and faced the loss of their houses. But they held out, nonetheless. As in past strikes, support for and participation in the labor action knew no boundaries of age or gender in mining households. Everyone in the household was expected to participate in the strike in every way he or she could. The mining households and communities policed themselves to ensure that there would be no "scabbing" in their neighborhoods. Men who crossed the picket line were subjected to brutal treatment by their neighbors and former coworkers. Often they were literally driven out of their homes and forced to take up residence elsewhere. The community would accept no excuse. People sometimes dug up the bodies of dead scabs and threw them out of community cemeteries. They did not want scab corpses to pollute "sacred" ground. The scabs' families would have to come and take away the caskets and find new burial grounds, sometimes on their own property. For the families of disinterred Catholics, this was a serious matter, because it meant that the body was going to have to be placed in unconsecrated ground.

By October 1902, with winter approaching, the situation in the coalfields began to turn ominous for large parts of the nation. Anthracite coal was the principal heating fuel of New England and the middle-Atlantic states and it was apparent to most people in the region that the strike's continuation into the winter would have dire consequences for their health and comfort. Public opinion, in great measure, was on the side of the strikers and several important civic organizations were calling for the nationalization of the mines. Even antiunion Republicans were warning President Theodore Roosevelt that something had to be done if the Republicans hoped to hold their congressional seats in the Northeast in the upcoming elections.

On October 11, 1902, President Theodore Roosevelt sent his secretary of war, Elihu Root, to have an unofficial visit with J. P. Morgan, the leading intransigent among the owners. They met on Morgan's yacht, *The Corsair,* in New York Harbor. Root, a former Wall Street corporate attorney, had a long and friendly relationship with Morgan. During this visit, the two men drafted a settlement proposal, which included an agreement on the part of the coal companies to the appointment of an arbitration commission by the president. The operators agreed to be bound by the commission's findings.

By October 21, the mineworkers had accepted the proposal for binding arbitration and Roosevelt had designated the members of the Anthracite Coal Strike Commission, as it came to be known. John Mitchell, the thirty-three-year-old UMWA president, chose the relatively young (forty-five-year-old) Clarence Darrow to represent the union and its members before the commission. While the hearings were taking place, the mineworkers returned to work. When the strike finally ended in October 1902, the miners had won minimal economic gains. The real gain they had achieved was organizational. They had demonstrated to themselves and to the companies that the myriad ethnic communities and workers in the various fields could come together around common goals and maintain solidarity no matter how brutalized they were by the coal operators and their police agents and public police forces.

Darrow already had established a reputation as a defender of labor. He had advised the Illinois governor John Peter Altgeld, who had pardoned the surviving men convicted in the Haymarket Square bombing; he had defended Eugene Debs, the labor leader who was prosecuted for contempt of court for events arising out of the 1894 Pullman strike; and he had successfully defended three members of the Woodcutters' Union in 1898, who had been charged under Wisconsin laws with conspiring to damage the business of a lumber company.

Darrow arrived in Scranton, where the hearings were to be held, in early November 1902. His friend, the prominent social reformer Henry Demarest Lloyd, joined him. Lloyd would work with him for the next four months of the hearings. In addition to Lloyd, Darrow had the assistance of four local attorneys who were familiar with the details of the mining business, and a large staff of paralegals. He also had a team of experts to help him marshal evidence and put together his case against the coal operators.

The coal operators put together a legal team consisting of twenty-four attorneys, who had an equally impressive support staff and an array of experts at their disposal. The parties arrayed against the mineworkers included three groups, each of which had overlapping, if not sometimes identical, interests: the major coal companies, which were the main producers and which transported their coal on affiliated railroad lines; the independent producers, who mined their own coal, sometimes on land leased from the majors, and who had their coal hauled on the major producers' railroads; and the nonunionized miners. All of these groups were united in their opposition to union representation of the mine workforce.

Attorneys for these groups launched a multifaceted attack on the union. The UMWA, it was maintained, had only an incidental interest in improving the conditions of the workers. Its real objective, the opposition attorneys claimed, was the seizure of the coal industry in order to serve the interests of the union's leadership. The principal weapons the union used to achieve its goals, according to the coal company attorneys, were violent intimidation and repression of anyone daring to oppose them.

The operators' attorneys further argued that the union, by means of the attempted regulation of production and hours of employment, was seeking to subvert and ultimately overthrow one of the most sacred principles upon which the United States was built: an individual's liberty to work where he wanted for as long as he wanted and for whom he wanted at whatever price he could negotiate for his labor.

The attorneys for the coal operators tried to make their case when they questioned union president John Mitchell. After having read his statement during the first day of the hearing and being led through his testimony by Darrow, the coal companies' attorneys subjected Mitchell to four full days of cross-examination (Phelan 1994). The operators' attorneys tried to make the following points: that if the workers got the wages they wanted, the poor would be hurt the most because the companies would have to charge more for coal; that the union was solely responsible for the violence that had marked the strike; and that the union was interfering with an individual's right to work by keeping nonunion miners off the job. The operators' attorneys made it seem that the very fabric of American civilization was under direct attack by the union.

When Mitchell was finished testifying, Darrow began calling his other witnesses, 244 people in all. The content of this testimony was varied: there was expert testimony on the operation of anthracite collieries, the nature of the anthracite industry, the health conditions created by working in the mines, and testimony from the local clergy on the everyday life in mining towns and mining patches. There was also testimony from the miners regarding working conditions, the skill requirements of their jobs, the number of days a year they worked, and how much coal could be produced by a miner in a day. One of the key complaints miners raised was that when they sent a loaded coal car to the surface, the amount of coal was estimated by a mine boss, rather than weighed. The miners believed that these estimates were inaccurate and cheated miners of just compensation for their work. Miners also gave testimony about the dangers of their work. Darrow called many injured and maimed workers to the stand. In response, company attorneys charged Darrow with producing a parade of horribles to win the sympathy of the commissioners.

Some of the most powerful testimony about the danger of mine work came from young boys who were employed in and around the mines. First, there was the testimony of the "breaker boys." At most anthracite mines, the coal, still in large chunks, was brought to the top of a five-or six-story structure outside of the mine called a "breaker." The coal was sent through a series of chutes where it was broken up into smaller sizes. Young boys straddled these chutes and pulled slate and other waste out as the coal moved through them.

The working conditions in the breakers were anything but safe and pleasant. The breakers were not heated, so in the winter, the boys, some as young as eight, worked under extremely cold conditions. In the summer they worked in intense heat. The noise from the coal moving through the chutes was deafening, and the air was filled with coal dust. You could not imagine a worse picture of an industrial hell than the inside of a breaker. Because the slate and coal has sharp edges, the boys' hands would be cut and bleeding. In the winter you could follow the paths they took home from work by the bloodstains in the snow. Many boys suffered the loss of fingers or parts of fingers picking slate from these rapidly flowing rivers of coal pouring down the chutes.

Another dangerous job in the mines was that of a "kipper." These were young boys who served as door tenders in the mines. They had to remain alert and open the door as 4 ton rail cars sped down slopes in the mines.

Following the testimony of the breaker boys, Darrow called the "mill girls," one as young as age eleven, to the stand. Her testimony showed that she worked a twelve-hour day and was compensated at the rate of 5 and 1/2 cents per hour. One of the main points the mineworkers stressed in their testimony was that their low wages forced them to keep their children out of school and, instead, send their children into the breakers and textile mills in order to make a living wage for their families. The mine owners' attorneys held that Mitchell and Darrow were being excessively soft hearted and sentimental about the use of child labor, which, as the companies saw it, was a good thing in most respects. Furthermore, they argued that miners were paid good wages and sent their children to work only out of greed, and not because of need.

On December 16, 1902, the presentation for the mineworkers' case was completed. The next day the nonunion mineworkers began pleading their case. On December 20, 1902, hearings concluded in Scranton; the commission re-opened to take evidence in Philadelphia on January 6, 1903. The nonunion miners continued with the presentation of their case against the union.

The major thrust of their evidence, that offered by the operators who followed them, was also that the union was a violent organization that did not warrant recognition as the bargaining agent for the mineworkers. While the operators presented some evidence purporting to show that miners' wages were adequate, that working conditions were safe and did not constitute an undo hazard to employees, and that the miners' overall living conditions were satisfactory, the main thrust of the operators' case was directed at the question of union recognition.

The question of whether or not the UMWA should be legally recognized was the most controversial and divisive at the hearings, as it had been since the time that the union had begun its organizing drive in the anthracite fields. The issue of union recognition had played a major part in the strike of 1902.

The key point that the operators and nonunion miners made against the union was that it was a violent organization and, as such, did not warrant organizational recognition. Attorneys for the operators and nonunion miners pointed to the violence that had been directed at the men who continued to work and at shopkeepers who continued to sell goods to non-striking miners.

Despite the importance of the question of union recognition, the commission's ability to consider this issue was unclear, given that, under the terms of the Morgan proposal that set up the hearings, the coal companies had agreed only to the arbitration of issues between the companies and their respective employees. They had refused to enter into arbitration with the employees' union. The problem now was that the mineworkers were asserting that, while they agreed that the union would not participate in the hearing, they had not agreed to remove recognition as one of the key issues to be resolved by the commission.

As it turned out, this procedural issue was not directly resolved at the hearings. But it was resolved, in a manner of speaking, in the commission's final report. In the report, the commission determined that the question of union recognition had not been before the commission because the first appearance of this demand had been at the first meeting of the commission on October 27, 1902. Therefore, the parties could not have understood it to be an issue when the agreement to arbitration was reached (U.S. Anthracite Coal Strike Commission 1903, 61).

The hearings closed on February 13, 1903. On March 21, 1903, the report to the president was released to the public. The report recommended that mineworkers get a 10 percent wage increase (they had asked for 20 percent), that the work day be reduced to nine hours (the miners had demanded an eight-hour day), that a board be established to resolve future disputes between the miners and operators, and that the union not be recognized because the issue had not been within the jurisdiction of the commission.

These gains for the miners were minimal. The wage increase did not cover the increase in the cost of living since 1900, when miners had received their last raises. While the nine-hour day was an increased benefit, the commission put no limit on the employment of children in and around the mines, and it rejected the miners' request that a system of weighing coal be implemented. Moreover, the union was not recognized as the employees' official bargaining agent.

The report was hardly a neutral document; in many places it explicitly embraced the view of the operators. For example, the commission observed,

> The Commission finds that the conditions of life of mineworkers outside the mines do not justify, to their full extent, the adverse criticisms made by their representatives. . . . It is true that the attention of the Commission was called to a few houses in which miners or mineworkers dwelt which were not fit to be called habitations of men, and there was testimony that others nearly as bad existed; but the disparity in human character is often manifested by a like disparity in homes and surroundings and this must not be lost sight of in considering the general conditions of the community. . . . When there is a question of the houses of recent immigrants . . . where they do not approach a proper standard, it is impossible to say how much choice and volition have to do with their inferiority. The homes and surroundings of the English-speaking miners and mineworkers are generally superior to those of the class just mentioned, and show an intelligent appreciation of the decencies of life and the ability to realize them. (ibid., 42–43)

Such statements by the commission amounted to a blanket condemnation of the mineworkers and their supposed habits of living. If the mineworkers lived in substandard conditions, it was a result of their own choices on how they wanted to live, rather than a reflection of the low wages and unstable employment offered by the operators. In a like manner the commission singled out immigrant workers for special attention. They were portrayed as not only lacking the character traits embodied in the native miners and mineworkers, but also for a general deficiency of culture that prevented them from wanting what were judged to be the decent things in life.

With respect to the claim that miners were not compensated appropriately for their work, the report stated, "We have already said that the personal element constantly enters into the case. The miner who by special aptitude of training knows how to economize powder and other supplies, and who is willing to devote two or three hours more a day than the average to his work, can and does make a larger income than his fellows who fail in these respects" (ibid., 49). Earlier, the report had noted, "Experience, natural capacity, aptitude for the work, individual industry, and habits of sobriety materially affect the amount that is earned" (ibid., 48).

As this report shows, the low wages earned by mineworkers were defined as a matter of individual character, industriousness, and talent and not the wage scale of the industry or how the industry operated. The commission also failed to make any recommendations about child labor other than that existing laws should be enforced more rigorously. Overall the commission report was more a vindication of the position of the operators than it was of the miners and certainly of the UMWA. At this time in the industry's history, this meant a vindication of the interests of the financier J. P. Morgan. In what was a conservative estimate, in 1901 the U.S. Industrial Commission estimated that Morgan interests directly controlled somewhere between two-thirds to three-fourths of all anthracite coal being shipped (Campbell 1938, 188–89). When his indirect control over the independent operators was added into this summary, this total increased to 90 to 95 percent (*United States v. Reading Company et al.* 1912).

Despite the commission's rulings, public opinion was clearly on the side of the mineworkers. The operators' position in the realm of public opinion was not helped by statements such as those made by George F. Baer, president of the Reading and a close confidant of J. P. Morgan. When the strikers' morale was at its lowest point, a Wilkes-Barre resident sent a letter to Baer asking him to settle the strike. Baer responded, "I beg of you not to be discouraged. The rights and interests of the laboring man will be protected and cared for not by the labor agitators but by the Christian men of property to whom God has given control of the property rights of the country and upon which the successful management of which so much depends" (quoted in the

Citizens' Voice, April 11, 1979). Baer's letter was made public, which prompted the *New York Times* to note, according to the *Citizens' Voice*: "A good many people think they superintend the earth, but not many have the egregious vanity to describe themselves as its managing directors" (ibid.).

In the following chapter, we describe further actions taken by Morgan interests and other forms of class conflict during the twentieth century before the Great Depression. As the century began, the mineworkers showed they were capable of forming a strong union that was able to harness their militancy and direct it toward more universal ends rather than toward parochial interests tied to religion, ethnicity, occupation in the mines, or geographic location. One of the major factors in giving workers a new type of collective consciousness was the penetration of corporate capital into the southern anthracite industry and its ascendance as the dominant form of production across all fields. A second factor was the integration of the industry under the control of Morgan's interests. This contributed to a standardization of wages and working conditions, and ended the cutthroat competition among miner operators and mineworkers across fields—a competition in which a strike in one field meant more work and, sometimes, higher wages for miners in other fields, only to have the wage increases and employment levels roll back when the strikes ended.

Chapter 5

The Post-1902 Strike Era

Morgan interests continued to try to break the UMWA following the 1902 strike. The attacks on the union came amid government investigations of the Morgan coal monopoly. In 1912 and, again, in 1920, the U.S. Supreme Court issued decisions that arose out of events and actions taking place in the anthracite coalfields in and prior to the time of the strike. In both cases, the Supreme Court ruled that the Morgan interests, using complicated and deceptive practices, flagrantly violated the Sherman Anti-Trust Act of 1890.

Labor unrest persisted after 1902 as workers struggled to defend their economic and social interests. In May 1910, there was a wildcat strike in the Pittston region, a mining community that was particularly militant and tumultuous. The strike was against the Pennsylvania Coal Company's operations. The strike began over a dispute regarding dockage rates for waste in cars of coal being brought to the surface. Soon the strike spread to other mines and much of the Pennsylvania Coal Company's operations were shut down. The UMWA, basically siding with the company by defining the strike as illegitimate, tried to force the mineworkers back to work and to refer their grievances to the Conciliation Board, as provided for in settlement of the Long Strike of 1902. The mineworkers refused. There were scattered riots in and around Pittston directed at the company and at mineworkers who tried to violate the picket lines. Most of the strikers were referred to as "foreigners" by the coal operators and their allies, which, at that time, meant eastern Europeans, central Europeans and/or Italians and Sicilians, but, generally, not second-or third-generation Irish, English, Welsh, or German immigrants. This was a way for the companies to try to discredit the strikers' cause and split the workforce. Police were called out on several occasions to suppress the strikers' violence against those trying to cross picket lines (*Record Year Book and Almanac* 1911, 29). This strike left a bitter aftertaste for mineworkers from Pittston, who felt that the UMWA had failed them by siding with the companies in and declaring the labor action a wildcat. This would not be the last time that the union and companies would ally themselves against the rank-and-file.

In 1912, there was another walkout of mineworkers. This included the three UMWA districts (Districts 1, 7, and 9) and was sanctioned by the union. The strike began at midnight on March 31 and stretched into the summer. The strike largely was over wages. In May that year, there were riots throughout the coalfields over how long it was taking the company to make what the miners regarded as an "acceptable" offer. These riots put pressure on the operators, who finally came up with an offer that, at first, was rejected by the mineworkers but later accepted (*Record Year Book and Almanac* 1913, 39).

Actions against the union were mitigated somewhat with the onset of the United States' participation in World War I. Entry into the war brought federal oversight of labor-management relations in the coal industry: something both sides agreed to because of the enormous profits the coal companies were making, and the high wages they were willing to pay, and the importance of anthracite coal to the war effort.

During the war, coal production reached all-time highs. In 1917, anthracite production reached 100.69 million tons, with 1/3 of it coming from Luzerne County alone (Keil 1988). The vast profits that were made were largely circulated back to corporate owners in Philadelphia, New York, and Baltimore. Enough was retained in the coalfields, however, to stimulate a minor economic boom, which resulted in a general rise in the quality of life for the mineworkers.

Despite the modest prosperity that came to mining families, the gain in living standards was largely relative. In comparison to other types of workers, mineworkers still were poorly paid (Lauck and Sydenstriker 1917). Researchers found that mineworkers had one of the lowest annual incomes of any blue-collar male heads of households, mineworkers' wages made up a smaller proportion of total family income than was the case for all other blue-collar male heads of households, and total family income of mineworkers' households was lower than that of almost all other blue-collar workers (ibid.). If these conditions prevailed during a period of high "prosperity," one can only imagine what life was like for the miners and their families during hard times, which were far more common than the good times of the war years.

After the war ended in 1918, anthracite production began to decline precipitously, as shown in Table 5.1.

As production fell, there was increased agitation in the minefields over declining wages and increasing unemployment, especially on the radical Left. During the early part of 1920, there was considerable interest on the part of

Table 5.1 Anthracite Production: 1895–1936

Year	*Millions of Tons*
1895	53.0
1917	100.7
1929	73.8
1931	59.6
1933	49.4
1934	57.4
1935	51.0
1936	54.8

Source: Ernst (1937).

the authorities in suppressing what they saw as rising support for Bolshevism in Luzerne County. According to the *Record Year Book and Almanac* for 1921, "It was known that there are many sympathizers with the strange Russian system in the County, among the foreign element." Furthermore, a number of meetings were broken up by the police. In January, officers of the Federal government made simultaneous raids in many cities in an effort to capture the most radical alien protestors and agitators. Forty arrests were made in Wilkes-Barre, Nanticoke, and a few other places in the upper part of the County (ibid., 11). The coalfields, however, figured only marginally in what came to be known as the "Palmer Raids," as several thousand people were arrested nationwide.

Nineteen twenty was a year of notable developments in the mining industry in the anthracite region. The miners had been operating under a wartime agreement that was independent of their contracts. This was set to expire in 1920 and be replaced by normal contracts. Both sides met in New York City on March 9 to begin contract negotiations. With negotiations stalled, the mineworkers began pressing for a strike. President Woodrow Wilson intervened and proposed that a commission be established to arbitrate the dispute. After several weeks of hearings, the commission submitted its findings and recommendations to the president in August. Two separate reports were given to Wilson: one, a majority report, was supported by the operators' and the public's representatives and the second was offered up by the mineworkers' representatives. The minority report favored acceding to the mineworkers' demands. The president decided to accept the majority report, which gave the mineworkers raises of 17 to 20 percent (they had asked for 30 percent) and ignored their other demands (*Record Year Book and Almanac* 1921, 25).

A group of dissident mineworkers, unsatisfied with the president's decision, met in Wilkes-Barre shortly after his decision was announced and voted to go on strike. The strike involved about 50 percent of all anthracite mineworkers in the region (ibid., 25). The strike crippled the industry and raised anxieties about labor actions spreading to other mineworkers and extending into the winter at a time when coal stocks were seriously depleted.

In the meantime, union officers and representatives of the operators met in New York City and signed an agreement based on what the president had accepted. The contract was for two years. For the first time in contract negotiations between the operators and the mineworkers, the name of the UMWA was on the contract. Prior to this, the wording on contracts had been such that the leaders of the three union districts had not been designated as representatives of the UMWA, but rather had been identified as "representatives of the anthracite workers." Thus the union achieved de facto formal recognition, something it had struggled to attain since the 1890s. But gaining recognition was costly for the union, in that it further compromised the trust the dissident mineworkers had in the organization. From the vantage point of the dissidents, the contract did not go far enough on behalf of the mineworkers.

The dissidents wrote a letter to President Wilson asking him to reconvene the commission, which he refused in a letter that accused the mineworkers of disloyalty. Upon the contract's signing, most of the dissident workers from the vicinities of Wilkes-Barre, Scranton, Pittston, and Carbondale returned to work. Strikers, however, remained out of work in the areas around Hazleton, Shamokin, and Pottsville. They did not return to work until after they had met with the UMWA's policy committee in September 1920, where they were pressured to end their strike and return to work. The dissidents also realized that it was futile to continue their walk out when the Wyoming and Lackawanna fields had returned.

Nineteen twenty-two saw another major strike in the anthracite coalfields. In January 1922, mineworkers began to assemble a list of demands in preparation for the expiration of their contract in March of the same year. They drew up a list of nineteen major demands. According to the *Record Year Book and Almanac*, their demands were as follows:

1. That the next contract be for a period not exceeding two years and that the making of individual agreements and contracts in the mining of coal shall be prohibited and where mechanical loading is done the committee and the company shall have authority to establish proper rates.
2. The contract wage scale shall be increased twenty percent and that all day men be granted an increase of $1.00 per day.
3. That a uniform wage scale be established so that the various occupations of like character at the several collieries shall command the same wage.
4. That the provisions of the eight hour day clause in the present agreement shall be applied to all persons working in or around the anthracite collieries coming under the jurisdiction of the UMWA regardless of the occupations, and that in the bringing of these employees under the eight hour day

their basis shall be arrived at in the same manner as the basis was arrived at in the case of pump men and engineers, plus the increase demanded in Section 2 of this document. And, further, that inside day laborers shall work on the basis of eight hours underground.

5. Time and half time for all over-time and double time for Sunday and holiday work.
6. The next contract shall contain a standard check-off procedure.
7. All dead work shall be paid for on a uniform consideration basis and that where more than one miner is employed they all should receive the same rate.
8. Payment for all sheet iron, props, timber, fore-polling, extra and abnormal shoveling and cribbing and where miners are prevented from working on account of supplies that they be accorded the opportunity of making a shift at some other work at the consideration rate.
9. In the settlement of grievances the aggrieved parties shall have the right to demand settlement on the basis of equity.
10. That a uniform rate of 17 cents per inch be paid for refuse in all kinds of mining up to 10 feet wide and a proportionate rate be applied for over 10 feet, with the understanding that this is to be a minimum rate not affecting higher rates that exist.
11. Where coal is paid for by the car it shall be changed and payment shall be made on the legal ton basis of 2,000 pounds and that dockage shall be eliminated.
12. That jack hammers be furnished free of charge to miners, including the power necessary to operate the machine.
13. A more liberal and satisfactory clause in the agreement covering the question of miners who encounter abnormal conditions in their working places.
14. That the wage schedules be brought up to date containing all new rates and occupations.
15. That carpenters and other tradesmen be paid the recognized standard rates existing in the region, which rate should not be less than 90 cents per hour.
16. That in retrenchment, the laying off of men and in the rehiring that seniority shall apply.
17. That employees of stripping contractors shall be brought under the general agreement on their present basis of wages and conditions plus the increase in Section 2 hereof.
18. That powder be delivered to the miners at their working places, or as convenient as possible to the working place.
19. That full eight hour opportunity be given to employees at collieries which have been working as a general schedule on a six or seven hour day, and that where eight hour opportunity is denied to those employes [sic] their wages shall be readjusted. (1923, 28)

Having formulated these demands and presented them to the operators, who rejected them out of hand, the miners struck on April 1, 1922. Six weeks later, on May 18, the operators responded to the miners' demands with a set of counter proposals.

> It is evident that the present economic situation demands a substantial decrease in wages if a normal production of anthracite coal is to continue and reasonably steady employment is to be provided. Therefore, in lieu of the wage program submitted by you, the operators propose an agreement embodying the following terms:
>
> a. Contract rates shall be decreased 18 percent below the rates established by the United States Anthracite Coal Commission in August, 1920.
> b. Day rates of men shall be reduced $1.20 per day or per shift below the rates established by the United States Anthracite Coal Commission in August, 1920.
> c. Day rates of boys shall be reduced 72 cents per day below the rates established by the United States Anthracite Coal Commission in August, 1920.
>
> This general wage structure represents an average decrease of approximately 21 percent and will therefore fully maintain the purchasing value of the wages as established by the United States Anthracite Coal Commission in August, 1920. It provides a minimum of 37 1/2 cents an hour for unskilled men employed outside the mines, with relatively higher rates for other occupations requiring skill and experience.
>
> With reference to the term of the agreement, the operators deplore the disturbance to business and the economic loss resulting from frequent controversies and suspensions. In order that this may be avoided, we propose a five-year contract, subject, however, to annual adjustments as to wage rates only.
>
> In case no agreement shall have been reached by March 1, in any year, the determination of proper wage rates shall be referred to a commission composed of five persons to be selected by the Presiding Judge of the United States Circuit Court of Appeals for the Third Judicial District. (ibid.)

Following the mine operators' offer, a long period of negotiations ensued, with neither side willing to compromise with respect to wage rates. The mineworkers continued to insist on raises and the operators on reductions. There matters stood until the intervention of President Warren G. Harding on July 9, 1922. On that date, President Harding, addressing miners' and operators' representatives, proposed a temporary compromise solution that entailed miners' returning to work at the old wage rates, the appointment of a board of arbitration, and the establishment of a commission to investigate the entire coal business in order to come up with recommendations for stabilizing the industry (ibid., 29). The operators agreed to arbitration, but the miners' representatives refused. They turned down the offer on the grounds that they generally lost in arbitrators' decisions, which typically favored the operators.

The strike finally was settled when the operators backed down on their request for wage reductions and agreed to extend the existing contract to September 1, 1923. The miners agreed to this plan on September 9, and work gradually resumed. The distribution of coal, however, was slowed down by a shop strike on the railroads, which reduced the availability of mine cars.

The mineworkers struck again in 1925. In the months and days leading up to the general strike, there were a number of small, scattered labor stoppages throughout the anthracite region (Kanarek 1974). Many, if not most, of these strikes were not sanctioned by the union, and union officials ordered the men back to work. In some cases, the strikers complied with union demands, but in others they did not. When the mineworkers refused to obey union officials, the national union officials sometimes suspended union charters, but these were only temporary disciplinary measures, as the union leadership knew that they would need these workers' loyalties for the larger strike that was likely to come toward the end of 1925, when the current contract expired.

The miners in 1925 held their tri-district conference in Scranton and drew up a list of demands to present to the operators: "A ten per cent wage increase for tonnage men, with an additional $1.00 a day for day workers, a two year contract and complete union recognition, described as another way of saying 'check-off' " (*Record Year Book and Almanac* 1926, 25). In addition to these, there was a longer list of miner demands, which included, among other things, improving the quality of company housing.

The miners' and operators' representatives began negotiating on July 9 in Atlantic City, New Jersey. Samuel D. Warriner, a representative of the operators, made a presentation stating that there could be no wage increase because company profits were being squeezed by the substitution of new home heating fuels, such as oil and natural gas. These had gained a foothold during and after the strike of 1922. John L. Lewis, the president of the UMWA, insisted that the mineworkers receive a raise on the order that the union was demanding, and he made it clear that the union would not agree to arbitration, given its past experiences with such procedures. By August 4, 1925, negotiations had broken down without having made progress toward any agreement.

Mineworkers struck in early September and the stalemate dragged on into October. Neither President Calvin Coolidge nor Governor Gifford Pinchot took steps to try to broker a settlement. Anxiety as to how people in the Northeast would fare in the face of a prolonged strike began to increase, and in response, newspapers and city officials across the region advised people to seek substitute fuels for heating their homes.

Finally, on November 28, Governor Pinchot submitted a plan to both sides to settle the strike:

1. That the price of coal shall not increase because of the suspension or the new agreement.
2. That the agreement shall extend at least five years.
3. The creation of a Board of Investigation and Award comprised of two men chosen by the miners, two by the operators, and three by these four.
4. The board to investigate whether the operators can pay increased wages without increasing the prince of coal.
5. The board to organize within thirty days and to make its report within six months.
6. Records of miners and operators to be open to the board but not to the public.
7. After making its award the board shall study methods of avoiding strikes and of increasing mine efficiency and operation.
8. Full recognition of the union not granted but operators shall honor voluntary written request of a miner to assign his present union dues, not exceeding $14 a year.
9. Working conditions not specifically mentioned to be referred back to representatives of miners and operators for settlement.
10. Board of Conciliation established under Roosevelt settlement of 1903, which never functioned, to proceed at once to equalize wages.
11. Suspension end at once. (ibid., 25–26)

On November 29, 1925, John L. Lewis accepted the governor's plan as a basis for an agreement to stop the strike and negotiate with the coal companies. The next day, the companies refused the governor's proposal, and the strike continued until February 1926, when it collapsed. The mineworkers, their collective and personal resources exhausted, suffered a major defeat in this strike when they returned to work on the company's terms.

In 1928, internal union turmoil broke out in the Pittston region of District 1 of the UMWA. The dissidents who believed that the UMWA had capitulated too easily to the companies had not been happy with the 1926 UMWA surrender. In addition, as economic conditions deteriorated in the mining industry, with production in 1928 at three-fourths the level of what it had been in 1917, the peak year of production, operators took advantage of hard times to put increasing pressure on workers and the union (Ernst 1937). Membership in the UMWA dropped dramatically throughout the coalfields. To preserve the integrity of the union, Lewis made a number of concessions to the coal operators (Aronowitz 1983). His willingness to compromise with the operators on issues that the mineworkers had fought hard to win led to high levels of discontent in the union, especially in District 1.

In 1928, members of a union local in Pittston rebelled against their local leadership, who were thought to be too close to the management, and elected a new set of officers. The union's District 1 president, Rinaldo Cappellini, then a Lewis ally, declared the election illegal, but the local's rank-and-file refused to accept his ruling. The local's insurrection led to the brutal murders of two militant, radical, progressive, dissident leaders, Tom Lillis and Alex Campbell (Rochester 1931). Campbell and Lillis had been long-time union activists. Both had been elected to office by rank-and-file dissidents.

Following these assassinations, Pittston exploded in violence. There were assaults, riots, and bombings by the dissident faction. John L. Lewis came into the region to try and restore calm. He blamed the insurrection on communist agitators, implying that Campbell and Lillis were part of a communist penetration of the local, and backed Cappellini and his associates. Eventually, Cappellini was forced to resign as District 1 president and was replaced by John Boylan, a Lewis ally and part of the union establishment. Boylan's appointment, however, did not quell the uprising, even though he and the board of directors of District 1 expelled the surviving officers of the dissident local.

As conditions deteriorated in the minefields with the onset of the Great Depression, the union's national officials became increasingly compliant in the face of the operators' demands even as the mineworkers became more radical in their demands. In July 1930, the union agreed to a five-year contract, something it had resisted in the past. The union also agreed to an extension of wages agreed to in the 1926 contract for a period of five years and seven months. In return for these two concessions, the operators agreed to a union dues check-off system, something for which the UMWA had been fighting for almost twenty-eight years (*Record Year Book and Almanac* 1931, 23). This was one more example of the union's organizational interests taking priority over rank-and-file interests.

The union also agreed to try to prevent wildcat strikes over the course of the agreement, to eliminate collective action by mineworkers to limit production to levels below what was possible at any given mine site, and to cooperate with producers in increasing the efficiency of miners' work, among other things. These stipulations turned over, in part, the policing of the labor process to the unions, rendering the union complicit in the subordination of workers in the mining industry. In addition, a permanent committee of twelve members, six from the union's side and six from the operators,' was created to oversee compliance with the new contract (ibid., 23).

Two years into the contract, which was due to expire on April 1, 1936, the mine operators asked the miners to accept a substantial cut in wages of up to 27 percent. The operators justified their request in terms of depressed conditions in the industry that had not been present when they initially signed the agreement. The union resisted taking a cut this large and the wage dispute was submitted to arbitration. The arbitrators were Frank Morrison, secretary of the American Federation of Labor (AFL), and a Washington attorney, George Rublee, who represented the operators' interests. They failed to reach an agreement (*Record Year Book and Almanac* 1934, 26). As a result of the stalemate, miners' wages were left unchanged.

The operators responded by making a second demand in April 1933. This time, they asked for a 35 percent wage reduction, and Rublee suggested adding a third arbitrator. A decision on the wage reduction request was delayed, and a proposed conference between the operators and the miners never met on the issue.

In March 1932, many locals in the union's District 1 voted to conduct an unauthorized strike. One of the main issues was work equalization, which Lewis refused to support and to which the companies were vehemently opposed. Under work equalization, no worker would be laid off and available work would be divided among the men. The strike shut down mining operations at several locations in Luzerne and Lackawanna Counties. Roving pickets moved from production point to production point in an attempt to expand the strike. They were not successful and the strike collapsed at the end of March 1932. That same month delegates from forty-four locals, representing 5,000 mineworkers in the Shamokin-Mahanoy cities area, went on strike against the Pennsylvania Iron and Coal Company. Again, work equalization was a major strike issue. In August 1933, more than 15,000 miners in the Panther Valley region struck the Lehigh Coal and Navigation Company over work equalization. This strike had the backing of local business and civic leaders as well as rank-and-file mineworkers (Miller and Sharpless 1985, 304–5). Despite the support for equalization, Lewis continued to oppose it.

The union took severe actions against the strikers and their leadership, suspending the leader of the strikers in District 1 from union membership for fifteen years and suspending rank-and-file mineworkers from West Wyoming, Exeter, Pittston, Larksville, Dupont, Duryea, Warrior Run, and Wilkes-Barre for shorter periods. All totaled, thirty-four mineworkers were suspended (*Record Year Book and Almanac* 1933, 21).

These suspensions did not end dissent in District 1. Indeed, it only seemed to intensify the tendency to rebel. In August 1933, matters came to a head when a new union was formed in District 1 called the United Anthracite Miners of Pennsylvania (UAMP) (*Record Year Book and Almanac* 1934, 25). Its members had grown disaffected by what they saw as collusion between the UMWA and the mine operators, as well as perceived corruption among some union officials. The new union had the support of communist organizers in the region (Nelson et al. 1981).

Decrying the lack of democracy in the selection of delegates to the July convention of the UMWA and the fact that rank-and-file miners were not allowed to express their grievances at the convention, a group of dissidents held a meeting in Scranton and organized the new union. Rinaldo Cappellini, who by this time had broken his alliance with Mitchell, was elected state president and Thomas Maloney, who had led the wildcat strike the year before and who was suspended from the UMWA for fifteen years for his actions, was elected president of District 1 of the UAMP.

The mine operators, working in cooperation with the UMWA, began acting immediately against members of the new union. Five-dozen members of the UAMP were fired from the Penn Anthracite Mining Company. When the company refused to rehire them, a strike was called against the company. The Hudson Coal Company fired a large number of workers for the same reason—they were alleged to be members of the new union. Like the Penn Anthracite Mining Company, the Hudson Coal Company was hit by a strike on September 24, 1933. A third company, the Glen Alden Coal Company, followed the lead of the two others and laid off a large number of men, all of whom, the company claimed, were members of the new union; a strike was declared against the company on October 17, 1933 (*Record Year Book and Almanac* 1934, 24–25).

Men who had stayed loyal to the UMWA refused to honor the new union's picket lines and tried to go to work, which led to violence as strikers and non-strikers fought each other at the picket lines. According to one report, "A reign of terror ensued. Scores of bombings and dynamiting occurred. Men who ignored the strike calls were attacked on their way to work and frequent clashes occurred between pickets and police" (ibid.).

Fearing even worse chaos, local bourgeoisie from Wilkes-Barre, Scranton, and Pittston traveled to Washington to meet with the head of the National Labor Board, Senator Robert F. Wagner of New York. The new union demanded federal intervention in the labor dispute and held a convention in Wilkes-Barre on October 30, 1933. They had invited a number of federal, state, and local government officials, none of whom showed up to the convention (ibid.). Partly in response to the boycott by public officials, the mineworkers proposed calling a general strike. Monsignor J. J. Curran, a local Roman Catholic clergyman who had been actively involved in helping to settle the 1902 strike, prevailed on the mineworkers to delay such an extreme reaction as a general strike until such time as invitations were extended to additional federal officials.

The mineworkers' meeting adjourned while they waited for responses from Washington. When responses were not forthcoming, the miners met again and, despite Curran's objection, called for a general strike on November 4, 1933, to take effect on November 6, 1933 (ibid.). On the day the general strike took effect, a senior technical adviser from the National Labor Relations Board in Washington, DC, arrived in Wilkes-Barre on a fact-finding mission.

He reported back to Senator Wagner, who, upon receiving and reading the report, told the miners that he would send a fact-finding mission to

District 1 if the miners would allow the collieries to reopen and allow non-striking workers to return to the pits without threats and intimidation. When Wagner offered assurances that men would be allowed to return to their former positions, regardless of the union in which they held membership, the dissident miners called a convention and voted to return to work (ibid.).

The strike was declared over on November 12, 1933. But when the men from the UAMP reported to work, many claimed that they were denied their former positions. Once again the situation in the coalfields turned dangerous. Curran rushed to Washington to ask that the commission promised by Senator Wagner be appointed immediately and begin its work without further delay. In response to the monsignor's appeal, the commission was appointed and began its investigation on November 20, 1933, hearing testimony from members of the UAMP and from members of the UMWA.

On November 18, 1933, in what could not have been worse timing from a public-relations perspective, John L. Lewis revoked the charter of Wannamie Local 400 of the UMWA, one of the first locals in District 1, charging that it was using UMWA funds to support dual unionism in its area. The officers of Local 400 denied the charges and proceeded to take legal action against Lewis (ibid., 26).

Given Lewis's action against Local 400 and the continuing attempts on the part of coal companies to repress the new union, UAMP men again began to demand a general strike by the dissident union. The UAMP District 1 President, Thomas Maloney, however, managed to have the men agree to postpone a strike until the commission finished its hearings.

At the same time that these hearings were being held, the operators proposed a "code of fair competition" based on the working conditions and wage scale in the existing contract. At the public hearings that were held in Washington, D.C., on November 17, 1933, the UMWA countered with a proposal for a reduction in the workweek to thirty hours and, in a turnaround for Lewis, the equalization of hours across all existing collieries. The operators opposed this demand and, thus, no "code of fair competition" was agreed to (ibid.).

Conflict among the UAMP on one side and the UMWA, and the operators on the other continued into 1934. On January 13, 1934, the UAMP called a general strike of its members and violent disorder ensued with "riots, dynamitings, shootings, and assaults. Numerous persons were injured and went to hospitals. Jail sentences were imposed. Property was damaged. Coal companies obtained injunctions which banned picketing" (*Record Year Book and Almanac* 1935, 26).

Much of the violence took place between miners who stayed loyal to the UMWA and those who supported the UAMP. The strike ended when James A. Gorman of Hazleton was appointed a federal umpire to hear the grievances of the UAMP miners, and the union agreed to his role. On February 9, workers agreed to return to the mines. The UAMP miners had filed some 1,037 grievances and complaints against the companies and the UMWA by the time Gorman began his hearings on April 3, 1934. Hearings were held for forty-five days in Wilkes-Barre and Scranton (ibid.).

On July 9, 1934, Gorman proposed a peace plan that featured a merger of the two unions. He reiterated his position on October 2, 1934, and directed both unions to accept his proposal. The UMWA agreed but not the UAMP (ibid.).

On October 9, Gorman issued a ruling that held that, "because of the existing agreement between operators and U.M.W. of A. and despite section 7a of NIRA (National Industrial Recovery Act) the operators could not recognize U.A.M. of P. committees" (ibid.). Coal companies could not be compelled to restore men who left their places during new union strikes, the umpire decreed, in one of his subsequent decisions, the majority of which were averse to the new union.

In mid-October, Thomas Maloney made an overture to the UMWA and the companies to peacefully resolve the situation in District 1. The companies rejected it, while the UMWA ignored it.

In convention, the UAMP voted down a proposal for another general strike, which some members had called for because of Gorman's rulings and the intransigence of both the operators and the UMWA. With a general strike avoided, however, neither the conflict nor the violence between the two mineworkers' factions came to an end. In Nanticoke, serious violence broke out when UMWA men refused to work with men from the UAMP.

Having been marginalized by Gorman's rulings and by the collaboration between the operators and the UMWA, the UAMP disbanded in 1935, and miners who were active in the insurgent union began applying for reinstatement in the UMWA (*Record Year Book and Almanac* 1937, 27). Former officers of the UAMP had to make their cases before an "international" commission, while rank-and-file members were reinstated by actions of the UMWA locals from which they had defected. In the interest of solidarity, virtually all of the former UAMP men were brought back into the UMWA.

On April 1, 1936, the UMWA contract with the operators expired. Despite several months of negotiations, little or no progress was made in reaching an agreement on a new contract. The miners had presented their demands to the operators' representatives on February 25 requesting the following: "(1) a thirty hour work week; (2) substantial wage increases; (3) work equalization; (4) the abolition of contract mining; and (5) a complete check-off system" (ibid.).

The operators countered with the following demands: "(1) a wage cut of 22.3 percent; (2) a plan for binding arbitration of disputes; (3) a five-year contract; (4) penalties for the union in the event of wildcat strikes; and (5) help from the UMWA in eliminating the massive theft of coal in the lower anthracite fields" (ibid.).

The theft of coal was an especially irksome problem for coal operators, especially the Reading. In the lower anthracite fields of Schuylkill, Northumberland, and Carbon Counties and, to a lesser degree, in the northern fields, unemployed miners, following the example of earlier generations, had invaded inactive coal lands and begun mining them under their own initiative. Independent breakers had emerged to process the coal, and an independent distribution system had developed to deliver coal to New York and Philadelphia, as well as to local markets. The mine operators could not gain control over the problem with their private security forces and Pennsylvania's Governor George H. Earle, a New Deal Democrat, refused to send in the state police to arrest the bootleg miners unless local authorities requested him to do so. None did, given that the local state structures were mostly under the control of Democrats who had been elected to office because of the backing of the coal miners or of Republicans who also had been elected because of mineworker support.

It has been estimated that at its highpoint there were twenty thousand men and boys engaged in bootleg mining, and that they were producing $35 million worth of coal a year (Miller and Sharpless 1985, 319). Many of these men were organized into a union that was developed with the assistance of the Communist Party in the coalfields. That the miners could engage in popular expropriation of coal lands with virtual impunity showed how weak the local bourgeoisie, coal operators, and state authorities had become in the face of miner militancy during the depression.

When Franklin D. Roosevelt ran for president he swept into office a large number of Democrats in Pennsylvania, giving workers a voice in government at all levels. A few Republicans in the coalfields were kept in office by mineworkers who could not bring themselves to switch parties.

Chapter 6

The Demise of Anthracite and the Rise of a Politically Organized Working Class

As the Great Depression deepened, the plight of the miners' worsened. Operators imposed deep wage cuts, placed thousands of men on short time, and laid off thousands of additional workers. The massive lay-offs and forced reductions in work time were a direct result of mine labor patterns that can be traced back at least to the Civil War, if not before. The mines' hiring practices involved building up a large reserve army of workers who were employed temporarily (when times were flush) but who worked part time or were laid off when times were hard. Mine owners saw this reserve army as "necessary" for a number of reasons. First, uncertainties of the coal market, which in large part depended on the nature of the weather in any given year made flexibility important. Harsh winters across the nation increased demand for anthracite coal, while mild winters decreased demand. Second, having a large reserve army was a defensive weapon within the cartel structure that organized production in the industry. If one of the cartel members decided to increase production, other producers had to be able to respond rapidly. The only way they could do that was to have at their disposal a large reserve army of skilled and semi-skilled workers they could rapidly deploy to swiftly ratchet up their own production. Third, the maintenance of a large reserve army of labor contributed to maintaining low wages and a modicum of labor discipline in the anthracite industry. It intensified the competition for work at the lowest possible wages and supplied a pool of possible strikebreakers.

The Great Depression deeply exacerbated the situation of mineworkers and their families. Newspaper reporters who went into the coalfields during those years wrote of seeing signs of widespread malnutrition and near starvation, and lack of adequate shelter or protective clothing during the winter months. They also wrote of children going to work full-time at the age of six to eight and competing against older men for the brutal job of breaker boys. These older men had once worked underground but were no longer fit enough to mine and now had taken "children's jobs." The work of the breaker boys and of their sisters, as seamstresses, silk mill workers, or housemaids, often meant the difference between economic survival and destitution for thousands of families in the coal regions. Looking at payroll records for a number of mining companies, the records show that the breaker boys often brought nothing home; instead, they helped pay the family debts to the company store, including the purchase of mining supplies for their fathers and older brothers.

Miners also took to spontaneous expropriation of company-owned coal, reviving an old tradition in the coalfields. They dug slope and drift mines in remote mountain areas and used the results to heat their own homes and cook their families' meals. Some set up illicit and illegal distribution systems. These wildcat, bootleg mines, or "rat holes" as they occasionally were called, were sometimes family-run operations. The son of one miner told how his father and the father's brothers-in-law and their sons opened several of these. The sons, all of whom were of school age, would take turns working in the mines. Some of the sons would go to school one year while others worked in the improvised mines. The next year, they would reverse roles. Their parents were determined that all of the boys and all their daughters would finish high school, something the parents had not done (Keil 1988).

Not all of the illegal mines were small-scale operations. Some were huge, numbering a hundred miners or more. Try as they might, the local and state police, as well as the company police forces, could not eradicate illegal mining. Often local authorities refused to take action against the illegal entrepreneurs, some of whom were their relatives or life-long friends. If arrests were made, many prosecutors would not press charges, and if charges were made, juries would not convict the mineworkers of theft of company coal. This left it up to the private police to take action against the mineworkers, which they did. They would drive miners off the land and dynamite the mines to make them unusable again. These actions did not deter the mineworkers who simply moved their operations deeper into the mountains where they were less easy to detect.

Bootleg mining was common throughout the anthracite coal region (Murphy and Murphy 1938). Mining families first began the modern form of this activity in 1929 by picking over the refuse dumps of coal owned by the companies. But the practice of seizing coal soon spread to other activities as mineworkers went into the hills and mountains to find outcroppings. The mineworkers sometimes sank tunnels 75 to 100 feet into the mountainsides. The miners also made deals with truckers to haul the coal to illegal breakers, which had opened to process the contraband coal. Whole families were involved in these activities, including women and children.

The "legitimate" coal operators finally appealed to the state for assistance to stop the illegal mining and the Pennsylvania governor established a "blue-ribbon" commission to examine the issue of popular expropriation. Like many such commissions, this one found that both the miners and mine operators had legitimate grievances. It made several recommendations, but none were ever implemented (Ernst 1937). Efforts to stop bootlegging generally were failures. There was widespread support in local mining communities for what the bootleggers were doing.

While, perhaps, stimulated in some degree by Communist and Socialist organizers, the popular expropriation of private coal reserves owes more to the mining traditions in the region than to clear-cut political ideology. Miners had engaged in the same behavior during the 1860s and 1870s in the depressions that followed the end of the Civil War. The expropriation was an example of pre-political, non-ideological behavior for most of the mineworkers. It ended as a mass practice when the depression drew to a close in the early 1940s with the onset of World War II, and a rise in the demand for anthracite. Mass expropriation was merely a strategy to mitigate the miners' economic misery, rather than a Communist-oriented mass radicalization as some feared and others hoped.

The organized appropriation of coal lands was not the only form of radical action taken in the coalfields during the 1930s. There was considerable support in the region for the formation of Communist-inspired "unemployed councils." An early organizer of the unemployed councils in the region was Steve Nelson, a Croatian immigrant who was born with the name Stjepan Mesaros. He was a Communist Party organizer who, along with his wife, Margaret, had been sent into the region to work with mineworkers in late 1930 (Miller and Sharpless 1985, 315–16). Nelson became an advocate for unemployment insurance for the miners, a defender of the bootleg miners, and an advocate of other "radical" causes (Nelson et al. 1981). He also supported the UAMP, seeing it as a militant alternative to the UMWA.

Under the Herbert Hoover administration, Nelson, and others like him, were continually harassed by federal, state, and local authorities. But this did not deter them. Conditions improved modestly for the radical Left organizers after Franklin D. Roosevelt was elected president. In the anthracite coalfields, the unemployed councils flourished, with over 20,000 members by mid-1931 (Ernst 1937, 316).

By the middle of the 1930s, the unemployed councils were firmly established in the anthracite regions. The Wilkes-Barre Council published a newspaper, the *Anthracite Unemployed Worker*, which had a circulation of five thousand. There also was a popular weekly radio broadcast sponsored by the Communist Party called the "Sunday Worker Radio Hour." Under Nelson's influence, the Communist Party in the region had their National

Miners Union cooperate with the UMWA, especially in the more radical areas of the mining districts. The party also supported the bootleg miners, helping to create the Bootleg Miners' Union (ibid., 319–320). For a detailed discussion of the National Miners Union and its activities in the anthracite region between 1928 and 1931 see Howard (2001, 99–124).

During its heyday, the local Communist Party supported the UAMP, as did the unemployed councils, which had heavy Communist participation. The unemployed councils also worked with the UAMP to prevent scabbing during strikes (Miller and Sharpless 1985, 320).

There were other radical movements on the Left besides the Communists, whose party membership in the coalfields numbered about two hundred in 1937 (Nelson et al. 1981, 174). Among these were the Anthracite Mine Workers of Pennsylvania (AMWP) and the Luzerne County Labor Party. They often took positions similar to the Communists in the region and worked in tandem with them on various issues (ibid., 163).

To return to the negotiations of 1935, little progress was made, as neither side proved willing to yield on its major bargaining positions. Just as it appeared that a strike would occur on May 1, negotiators, facing pressures from Assistant Secretary of Labor Edward F. McGrady, agreed to extend negotiations until an agreement could be reached. McGrady presided over around-the-clock negotiations for six days until an agreement between the two sides was reached. The agreement was to be retroactive to May 1, 1935, and was to last until April 30, 1938 (*Record Year Book and Almanac* 1937, 26).

The wage rates fixed in the contract of February 17 were to be extended until May 1, 1937, when the miners were to get a seven-hour day, five-day workweek (35 hours per week) with the same pay they had previously received for eight hours per day. The contract also provided for a form of work equalization and a check-off system for union dues. In addition, the contract called for penalties for the union in the event of wildcat strikes and charged union officials with the responsibility for administering their side of the contract, thereby further incorporating the union into the structures controlling the mineworkers' labor process (ibid.).

A memorandum of understanding, which was not part of the contract, was written. It provided for the establishment of a joint commission made up of six miners' representatives and six operators to examine a host of issues plaguing the industry, including health and safety conditions for miners and bootleg mining (ibid.).

The contract was approved by an overwhelming majority of mineworkers on July 19, and signed on July 20. Some of the independent companies initially refused to sign the agreement, and their workers went out on strike. By late August, however, the independents had capitulated to the union agreement and ended the strikes.

During 1936, the operators created Anthracite Industries, Inc., a marketing firm, and pledged $3,000,000 to support the new company. The goal of the company was to win new markets and restore old ones that the operators believed had been lost due to labor instability, which, in the operators' view, had forced consumers to seek out alternative fuels such as oil and natural gas. It also was during this time, however, that the industry began falling on even harder times because of the withdrawal of capital investment and the moving of this investment into fuel alternatives to anthracite coal.

This investment in alternative fuels amounted to the first wave of deindustrialization experienced by the anthracite region, as it involved the export of jobs by coal producers. Producers began shifting production from northeastern Pennsylvania to other mining regions, where wages were much cheaper and other mining costs were lower.

As general economic conditions deteriorated nationally during the Great Depression, conditions deteriorated much more extensively in the anthracite fields. Coal operators took advantage of "hard times" to seriously weaken the UMWA and even to break the union where possible by putting into effect non-contract work rules and wage rates. Union membership dropped sharply. To preserve the union's existence, John L. Lewis made a number of significant concessions to coal operators (Aronowitz 1983). Not all locals were willing to accept these concessions. Lewis's willingness to offer concessions generated a great deal of discontent throughout the anthracite fields.

Lewis and the UMWA establishment ultimately triumphed over the dissidents by making some concessions to the dissidents' demands and, being no stranger to internal repression, forcing other dissidents out of office or co-opting them. This strategy, however, left many of the issues that had spawned the rebellions against the UMWA unresolved. The lack of a resolution of the various locals' grievances led to more violence in the 1930s.

As the industry continued to lose market share, various proposals arose as to whether the federal or state governments ought to step in to regulate the industry. At one point in 1937, the Commonwealth of Pennsylvania proposed taking over control of the mines, but this was rejected by the legislature, as was federal legislation designed to stabilize the industry (*Record Year Book and Almanac* 1938, 42).

The small oligarchy of companies that controlled the industry blamed the mineworkers and their willingness to strike for the continuing decline of the market for anthracite coal. What the oligarchy, however, overlooked was the price for coal compared to alternative fuels. A good part of the price differential between these fuels and anthracite were not a function of the cost of labor but, rather, were more immediately a function of the costs of coal freight rates. The Pennsylvania Public Utilities Commission, maintaining that freight rates for coal were unreasonable, proposed a 20–37 percent reduction in rates from the Wyoming, Schuylkill, and Lehigh fields. In the meantime, coal haulers asked for a 15 cent a ton increase, which the Public Utility Commission rejected, despite the acceptance of the rate by the Interstate Commerce Commission (ibid., 43).

By 1938, production in the anthracite mining industry had declined to approximately 46 million tons. The industry had become a shell of what it once was. While almost 97,000 men were still employed in anthracite production, this was only a fraction of what the employment had been a decade before. The industry seemingly was in its death throes.

In 1940, the operators and the miners negotiated a two-year contract in which miners were granted a 7 1/2 percent pay increase until October 1 and a 10 percent increase thereafter. The agreement was signed on June 20 after a one-day strike had shut down the mines in May. The mineworkers, for the first time, obtained ten days of paid vacation at the rate of $20 or $2.00 per day. The contract was approved by an overwhelming vote in a referendum of all mineworkers, rather than in a convention, which had been the process previously. Some locals voted against the contract, mainly because of provisions that permitted the UMWA to charge additional assessments that would be collected by the coal companies (*Record Year Book and Almanac* 1942, 51).

Internal conflict within the UMWA over the contract provision led to a month-long strike in District 7 in the southern coalfields. It threatened to spread into Luzerne County as roving picket lines made up of men from the district showed up to force walkouts at Luzerne collieries. Some mines in the county were shut down for a few days, but eventually the men in District 1 decided not to act in solidarity with the men in District 7. As a result the strike in District 7 collapsed due to worker fragmentation.

President Lewis of the UMWA, who opposed the actions of the District 7 men, took his revenge on the District 7 strikers by having the international union take over the management of the district. All of the district officers were removed and replaced by Lewis's appointees.

An Anthracite Allocation Plan, which allotted production quotas to different companies, was made into state law in 1941, in effect legalizing an oligopoly. As the nation geared up for war in 1941, profits in the coal industry soared as demand increased and the number of labor actions declined dramatically in 1941 (ibid., 52–53).

With the advent of World War II, and the consequent increases in demand for anthracite, the pool of surplus labor was redeployed as active labor. Indeed, there even were significant labor shortages in the coalfields, as men went into the military or moved out of the region to take jobs in rapidly expanding defense industries. Chronic labor shortages caused some cities, such as Philadelphia and New York, to face serious shortfalls in the supply of coal that they needed to heat homes and apartments and provide fuel for factories.

The War Labor Board, which had taken control over labor issues during the course of the war, granted the anthracite mineworkers substantial wage increases in 1944 and managed to keep strikes in Luzerne and Lackawanna Counties to a minimum. Strikes were more common in the southern anthracite coalfields during 1944, where the UMWA was weaker in maintaining labor discipline. One estimate suggests that more than 10,500 miners were on strike for one reason or another in August 1944 (*Record Year Book and Almanac* 1945, 52).

With the end of the war, demand for anthracite plummeted. By this time, the country had already substantially shifted to new ways of heating homes and providing fuel for industry. Hard times returned to the anthracite region. Production levels fell, unemployment rose, and corporate owners accelerated their withdrawal of capital from the industry and the region. Between 1950 and 1955, 20,600 jobs were lost in the industry; from 1956 to 1978, another significant drop in employment occurred, with most of the shrinkage occurring between 1956 and 1960 (Pennsylvania Bureau of Statistics and Planning 1974). With the loss of jobs, significant out-migration from the region took place, as workers left to seek employment along the Great Lakes coast, in northern New Jersey, and in Philadelphia, all emerging centers of America's post-World War II prosperity. Many of these out-migrants recognized coal's downturn was not cyclical this time, but was a permanent change.

On an unusually warm night in January 1959, deep mining in the Wilkes-Barre region was brought to a fitting, bloody close, when the Susquehanna River broke through the roof of the Knox Mine just south of Pittston and all deep mines south of that point were flooded. Twelve men were killed, and millions of gallons of water poured into the mine before the cave-in could be repaired. By the time the roof was sealed, the flooding had killed off all of the deep mines in the Wyoming Valley.

By the time of the Knox cave-in, there was little left of the coal industry in the anthracite regions. Few mines were working, and of those that were, fewer still provided full-time employment. Many of the corporate producers had declared bankruptcy and shut down their mines, walking away from thousands of acres of scarred land. Others sold their assets or leased them to under-financed entrepreneurs. These undercapitalized operators produced for an ever shrinking, localized market. In order to make profits, the new petty producers often resorted to unsafe mining practices—such as robbing pillars (cutting into columns of coal that supported the roof a mine) or mining too close to the surface. It was just such actions that led to the Knox disaster.

The investigations following the aftermath of the Knox cave-in showed the sorry state into which the industry had descended and the corruption that had crept into corporate mining practices and the once-mighty UMWA, including penetration of the industry by organized crime (Wolensky et al. 1999). In the 1930s, when labor relations were especially difficult, demand was low, and bootlegging of coal was prevalent; the larger mining companies had turned to leasing their mineral rights to independent entrepreneurs who, in return, pledged to sell all of the coal they mined to the larger corporations. The entrepreneurs bore all the risks of mining, including conflict with labor, while the large corporations enjoyed substantial profits with minimal risks.

The leasing system put downward pressures on wages and working conditions, as potential contractors tried to produce the highest possible bid while still allowing themselves room for profits. This could not be done without violating basic union labor laws and wage standards. In turn, this situation led to the corruption of local union officials, who were willing, for a price, to ignore violations of the labor contract and insure labor peace at the point of production. The larger companies awarding the leases were themselves not immune to corruption, which was especially clear in the Knox case. There the Knox Corporation's lease awarder, the Pennsylvania Coal Company, allegedly gave Knox permission to mine pillars and to move mining operations very close to the bottom of the river, in areas where there were good reasons to suspect that mine operations might be unsafe. Presently, the coalfields contain vast anthracite deposits, yet little mining is being done. Production there has come largely from surface operations.

During the century in which anthracite mining held sway in northeastern Pennsylvania, coal operators never achieved the levels of stability in the production system that would have permitted operators to integrate their workers into a relatively smooth "social structure of accumulation," in which workers took for granted the capitalist mode of production with its chronic crises, periods of mass unemployment, and closer and closer regulation of work in a more finely carved division of labor (Gordon et al. 1982, 10).

One of the main obstacles impeding the development of an effective, smoothly functioning social structure of accumulation was the intense competition that plagued the industry throughout most of its history—a system of competition that took place within and across fields. While internal competition was finally tamed to a modest extent in the 1880s and 1890s, the anthracite producers soon found themselves in competition with the bituminous mines in western Pennsylvania, West Virginia, Kentucky, and the rest of southern Appalachia, stretching all the way into northern Alabama. Never able to achieve a monopoly for themselves or, even, a stable oligopoly through long-lasting cartels, the mine operators continually squeezed the workers to gain profits.

As a result, mineworkers and their families endured and struggled against some of the worst forms of economic, social, and political exploitation imaginable. Not only did mineworkers have to experience low wages and irregular employment, but they suffered other indignities as well. Companies deployed their private police forces to control the workers and their communities. The Coal and Iron Police had been granted sweeping powers of arrest by the Pennsylvania legislature in the 1860s. The companies used private detective agencies such as the Pinkertons to spy on workers and their families and organizations. The companies also organized private vigilante groups to repress worker rebellions. They blacklisted dissident workers and fanned ethnic animosities among the working-class populations in the region.

Miners who worked and lived in the company towns were the worst off. They experienced a double exploitation as producers *and* as consumers. As was the case elsewhere in the coal industry, miners in company towns in the anthracite fields were often paid in company scrip (Corbin 1981). The scrip was redeemable only at the company store, where men not only had to buy their work supplies but their household supplies as well. The company stores were known for selling shoddy, inferior goods at inflated prices because they had a captive consumer base, which dared not complain and had no cash to buy their goods elsewhere.

The companies also ran the local elementary schools and supplied the ministers and priests for religious services. Given that the clergy were bought by the companies, Christian "teachings" were often harnessed to the companies' needs for profits. The companies also provided medical care and determined who were suffering from work-related injuries and deserved compensation versus who were "malingerers" who deserved no compensation.

In company towns, houses were built to substandard levels, sanitary conditions were abysmal, and diseases were pandemic. In addition, residents of these towns were subjected to far more intensive and extensive social control systems than were miners who lived in public municipalities. In the company towns, there was little social space in which miners and their families could develop a private civil society and private, autonomous social relations. Public and semi-public places, such as taverns and churches, were filled with company spies who reported on mineworkers' conversations. If the talk became too hostile to the company or too pro-union, a worker could be fired from his job, evicted from his home, and blacklisted in the industry, making it impossible for him to find another job in the regional coal industry.

During their long years of struggle to build a union, the miners not only had to contend with the power of the private police forces, but also with the policing agents of the state. In the nineteenth century, this typically involved calling up various militias from around the state and sending them into the coalfields to "restore" order by helping the company to break the strike. After 1905, the state police performed this function and did so with the same level of brutality as had the militias. We interviewed an elderly man who recalled a strike when he was a young boy, sometime in the late 1920s or early 1930s, when a group of miners had gathered in his yard (his father held some type of minor union office). The state police, mounted on horses, came through the family's wooden picket fence and began beating the men in the yard. Several were injured severely. Some years later, the man, his father, and three of his brothers, spotted the corporal who had been in charge of the state police unit at a holiday picnic in the region. They waited until he was leaving, grabbed him, and beat him badly for his actions ten years earlier.

Because of the close intertwining of politics and capital, and the willingness of capital to use the state to crush workers' actions, all of the mining unions, stretching back to the WBA, and forward to the UMWA, were politically activist. While they could capture a few seats in the Pennsylvania General Assembly, the miners' real political success was on the local level. There, they managed to claim quite a few town council seats in the suburban mining towns surrounding the larger cities, elect several mayors, and gain control of local police departments. They also won control over other local institutions, such as health commissions and public schools. The latter were especially important in inculcating working-class values in younger children at the same time that they equipped students with fundamental skills such as reading, writing, and basic numeracy.

The unions generally worked through the Democratic Party in the region, after their brief flirtation with the Greenback-Labor Party in the 1800s and a Labor Party in the 1930s. After the 1850s, the local bourgeoisie largely worked through the Republican Party. Prior to that in the 1840s, they supported the Whigs and some were involved with the American Party, an anti-Catholic, and anti-immigrant organization. The United Order of American Mechanics and the Junior United Order of American Mechanics, made up of Protestant workers and small businessmen, and similar organizations also supported the American Party. These groups went so far as to require that their members do no business with Catholic merchants nor work with or hire Catholics. These organizations truly believed that Catholic immigrants were part of a Jesuit "plot" to seize power in the United States and convert the nation into a Catholic state. While such nativism was especially strong among the artisan strata, the local labor aristocracy in the region, it was truly a cross-class movement, binding the bourgeoisie, skilled workers, unskilled workers, and farmers into a political bloc that stood opposed to first the Irish and, later, newer immigrant groups from central, eastern, and southern Europe.

Eventually the workers took control of the cities of Wilkes-Barre and Scranton and most offices in Luzerne and Lackawanna Counties. In doing so, they broke the structure of accumulation that had been in place for almost seventy-five years—a structure based on an economic and political class alliance between the local bourgeoisie and extra-local corporate interests, and on brutal economic and political exploitation. This structure of accumulation, while it lasted, redefined northeastern Pennsylvania economically, politically, and spatially, all in the name of extracting as much profit as possible from the region's natural energy resources.

The structure of accumulation, eventually, was dismantled in the coalfields through the concerted political activity of the working class. On an economic level, at the beginning of the twentieth century the historical competitive anarchy in the anthracite industry was being held in check by the hegemony of Morgan banking interests out of New York City. J. P. Morgan's interests had put together a strong oligopoly of producers and used his financial weight to police the system. But just at the moment when the anthracite industry had reached its peak integration, it began to face intensifying energy competition. The full effects of the competitive pressures were delayed because of World War I, but after the end of the war, they became immediate influences on the direction that the anthracite industry would take for the remainder of the century.

By the beginning of the twentieth century, the class-based political rule of the local bourgeoisie was being challenged on all fronts. For example, in 1906, Thomas D. Nicholls, the radical and militant leader of District 1 of the UMWA, won election to the U.S. Congress as a Democrat, breaking, for the first time, the monopoly that the bourgeoisie had held on the region's congressional seat. Nicholls served three consecutive terms before being replaced by another union leader, John J. Casey, president of the union of Plumbers and Steamfitters. Casey served six terms in Congress. In 1944, Daniel J. Flood won the seat. Flood was a local attorney with family and personal ties to the UMWA. He lost in the next election in 1946, but won again in 1948 and was reelected in 1950. He lost the 1952 election that swept Eisenhower into office but won and held the seat from 1954 until 1980, when he was forced to resign to avoid a second trial on federal corruption charges. Flood pled guilty to several minor charges that one local political wag described as akin to having not parked his bicycle straight (Keil 1988). During his tenure in office, Flood was a flamboyant character who affected the look of a Mississippi riverboat gambler. He wore pastel shirts, a string tie, and waxed handlebar mustaches, and always drove a Cadillac convertible. It seemed that a man presenting himself in this manner would be "out of touch" with his constituency of hard coal miners and their families. But Flood was not. His constituency loved him, and not only for the federal patronage that he brought to the Wyoming Valley, but also because of his outlandish dress and behavior. Flood and his family also had historical personal and professional ties to the UMWA and other unions in his congressional district.

When Flood resigned his seat in Congress he was replaced by State Representative Rafael Musto. Musto served out Flood's term but was defeated in the 1980 election. Musto also had close ties to organized labor especially the International Ladies' Garment Workers' Union (ILGWU) in the Pittston region. Musto lost to James Nelligan, who had extensive federal government experience and proved to be an effective campaigner. Musto was elected to the state senate in 1983. He served until he was indicted and convicted of corruption. Nelligan served one term in Congress and was defeated in 1982 by Frank Harrison. Harrison was an attorney, having graduated from Harvard Law School and a college professor at Trinity University. He lost in the Democratic primary to Paul Kanjorski in 1984. Kanjorski won the 1984 election and held the seat until a loss in the 2010 elections to an anti-immigration candidate.

The Democrat-labor alliance was successful not only in winning congressional elections but also in claiming a number of local and state seats throughout the coalfields. Between 1900 and 1920, it succeeded in electing several pro-labor candidates to the state's General Assembly (Keil 1988).

The political fortunes of labor and the Democrats dramatically changed for the worse in the 1920 elections. Following national trends of that year, Democrats were swept out of almost all of the offices they controlled at the county and state levels in the coal region. By the end of the decade, there were few Democrats in the coal counties' state legislative delegation. For example, in 1926, there was only one Democrat in the state house from Luzerne County and only one Democrat in the state senate from Luzerne County. By 1927, both of these Democrats were out of office.

Between 1927 and 1936, the Democrats never held more than two of Luzerne County's eight seats in the Pennsylvania house and did not win an election to the state senate. Local political dynamics changed in 1936 when the Democrats, putting together a coalition of workers and the descendants of immigrants from Ireland and southern, central, and eastern Europe, swept back into state and local political office on the strength of President Franklin D. Roosevelt's landslide victory.

In 1942, with the onset of World War II having ended the depression in the coal industry, Republicans regained strength in the coalfields. Their hegemony lasted a decade. As an example of the political imbalance in the region, in 1946, at the end of World War II, Republicans in Luzerne County had approximately 135,000 registered voters to the Democrats' approximate 52,000, a ratio of 2.6 to 1. A crucial turning point took place in the region's politics in 1959. After that year, the Democrats became the major party in Luzerne County and in Wilkes-Barre city. This shift was helped, but not defined, by the presidential candidacy of John F. Kennedy. His 1960 campaign shifted a large number of Catholic voters (who were a substantial minority in the county) into the Democratic ranks, where they remained until the candidacy of Ronald Reagan, who attracted large numbers of working-class, Catholic votes, in the region.

But, in 1978 the Democrats had a voting edge in Luzerne County of 108,948 to the Republicans' 65,334. The Democrats had sizable majorities in all of the county's major cities. For example, in Wilkes-Barre, there were 16,208 registered Democrats and 7,637 registered Republicans. The Democrats had almost 5,000 registered voters in Pittston and the Republicans just 744. In Hazleton, the Democrats had slightly more than 9,000 registered voters, with the Republicans almost 5,500. In Nanticoke, the Democrats had a greater than 5-to-2 majority of registered voters (*Pennsylvania Manual* 1979, 776–77). Republicans, on the other hand, held majorities in a few of the suburbs immediately adjacent to Wilkes-Barre and in the semi-rural fringe and in rural communities in Luzerne County. In Northumberland County, Democrats out-registered Republicans in Sunbury, but not in Shamokin (ibid., 785) and in Pottsville in Schuylkill County Democrats predominated over Republicans (ibid., 788).

The period of political instability from the 1930s to the mid-1970s, with alternating periods of Democratic and Republican local hegemony, was one of struggle for the working-class and immigrant groups as they fought to build a political machine to break the monopolies of the old, local bourgeoisie and their class allies. It expressed the rough balance of forces between the local bourgeoisie and their class allies and the working classes and their allies to gain hegemony over Luzerne County's political life. By the middle of the 1970s, the workers had conquered local political power and pushed the bourgeoisie and their allies to the periphery of local politics, where they remain today. The local bourgeoisie no longer are totally interlinked with the local political elites, who are more closely connected to working people and the trade union movement.

The Democratic Party could not have emerged as the dominant political party in several of the mining counties had it not been for its close ties with labor and with immigrant populations. The Democratic Party was virtually fused with the trade unions and the communities of descendants of former immigration streams; just as Republicans once had thrived from a fusion of the local bourgeoisie, smaller entrepreneurs, and native elements of the population.

In 1980, the ratio of registered Democrats to Republicans in Carbon County was 1.26 to 1; in Lackawanna County, it was 2.10 to 1; and in Luzerne County, 1.59 to 1. In both Northumberland and Schuylkill Counties, however, Republicans dominated, with registered Democrats to Republican ratios of 0.86 to1 and 0.69 to1 respectively (*Pennsylvania Almanac* 1980).

The same pattern occurred in 1984, with registered Democrats outnumbering registered Republicans 1.28 to 1 in Carbon County, 2.23 to 1 in Lackawanna, and 1.6 to 1 in Luzerne County. In Northumberland and Schuylkill Counties, the ratios were reversed, standing at 0.96 to 1 and 0.71 to 1, respectively (*Pennsylvania Almanac* 1984).

By 1988, there was a shift in the pattern. In Carbon, Lackawanna, and Luzerne Counties registered Democrats continued to outnumber registered Republicans by 1.23 to 1, 2.3 to 1, and 1.52 to 1, respectively. Democrats, unlike in 1980 and 1984, predominated in Northumberland County, while Republicans continued to maintain their dominance in Schuylkill County. Northumberland County's ratio was 1.03 to 1, for a slight Democratic advantage, while the ratio was 0.71 to 1 in favor of Republicans in Schuylkill County, just what it had been in 1980 (*Pennsylvania Almanac* 1988).

In 1992, the pattern was the same as in 1988. Democrats exceeded Republicans in Carbon, Lackawanna, Luzerne, and Northumberland Counties by ratios of 1.29 to 1, 2.2 to 1, 1.61 to 1, and 1.05 to 1. Only in Schuylkill County were Republicans in the majority, with a ratio of 0.76 Democrats to 1 Republican (*Pennsylvania Almanac* 1992). The same pattern held for 1996 and 2000. In 1996, the ratios for Carbon, Lackawanna, Luzerne, Northumberland, and Schuylkill Counties were, respectively, 1.24 to 1, 2.22 to 1, 1.60 to 1, 1.03 to 1, and 0.75 Democrats to 1 Republican. In 2000, the ratios for these same counties were 1.22 to 1, 2.03 to 1, 1.61 to1, 1.02 to 1, and 0.75 Democrats to 1 Republican (*Pennsylvania Almanac* 1996, 2000).

Democrats, in 2004, maintained their hegemony in their traditional areas of support: Carbon County (1.22 to 1), Lackawanna County (2.06 to 1), and Luzerne County (1.59 to 1). Schuylkill County remained Republican, with a ratio of 0.77 to 1, and Northumberland County reverted to the Republicans, with a ratio of 0.92 to 1 (*Pennsylvania Almanac* 2004). This structure was reproduced in 2008, with Carbon, Lackawanna Counties having ratios of 1.36 to 1, 2.39 to 1, and 1.82 to 1.

In three counties (Carbon, Luzerne, and Lackawanna), 2008 was a year when Democrats most strongly outnumbered Republicans over the time period considered here. Northumberland and Schuylkill Counties had respective ratios of 0.97 to 1 and 0.89 to 1 (*Pennsylvania Almanac* 2008).

The success of Democrats in Carbon, Lackawanna, and Luzerne Counties was a result of the Democratic Party's linking up with the labor movement and with the various ethnic communities that comprised the three counties. Forging the alliances with ethnic communities, particularly the Irish, however, contributed to Schuylkill County's and, to a lesser degree, Northumberland County's being Republican. Historically in these two counties, the citizenry, including the various components of the working classes, tended to see the Democratic Party as an Irish and Catholic political instrument, something not acceptable to workers who had a historical memory of conflict with Irish in the mining industry and who were leery about the Irish presence in the commonwealth's Democratic Party.

Direct political activism on the part of trade unions and, in particular, their alliance with the Democrats in three out of the five former anthracite counties is a legacy of the mining era. The old Workers' Benevolent Association, the Miners' and Laborers' Benevolent Association, the Knights of Labor, and the United Mineworkers were all politically active in the surrounding region. All had recognized that the miners' historical patterns of local terrorism and other forms of pre-political action were insufficient for advancing, let alone protecting, their members' interests. In the view of the mine unions, which dominated by sheer size alone the local trade union movement, direct economic actions against employers had to be coupled with direct and sustained political activism.

Combining economic and political action, the region's unions, especially the UMWA, brought major improvements to workers' and workers' families' lives. Wages were advanced, conditions of job stability were improved, safety and health improvements at work were made, and the most brutal forms of state-sponsored or state-protected violence against mineworkers were brought to a close. Even the notorious Coal and Iron Police were forced to disband in the 1930s, with the Pennsylvania state police picking up most of its law enforcement functions. While the state police could be and often were as brutal in repressing worker dissent as the Coal and Iron Police had been, eventually, as unions gained political power, the repression by the state police was considerably tempered and dampened.

Political activism was not the only legacy the mining industry left to the region's working classes. The UMWA also taught the local working classes that unionism would be successful to the degree that unions developed high levels of respect for and cooperation with each other. The UMWA showed this cooperation by helping to found and sustain other local unions and, eventually, by working to establish the Congress of Industrial Organizations (CIO).

The mining industry left the region with a distinctive social structure and local culture. It was a social structure that was heavily working class and bore a high commitment to the culture and values of unionism, making the region a union community in the fullest possible meaning of that term. Unions were and are vested with a popular moral value, not only as a means to obtain higher wages and better working conditions but also as a means to achieve dignity in the eyes of other people in the community and in the eyes of employers as well.

Building such a union community was no easy task. It required forging strong bonds within and among religio-ethnic groups. Contrary to what some sociologists have argued and what large parts of the general public believe, as we noted previously, few of the immigrant groups coming to the region arrived with well-developed national/ethnic consciousness of themselves as coherent groups. Nor did they arrive with high levels of structural unity. They were drawn from many different regions, and provinces, and villages in the "old country" and lacked the central thing that bound these communities together—social trust. They would achieve this trust, a collective identity, and structural integration after immigration and settlement in mining communities through participation in churches, taverns, and mutual benefit societies, and by sharing space and developing formal and informal institutional structures in their shared space (see, for example, Keil 1988; Yancey et al. 1976; Reich 1977; Oppenheimer 1974, and Cox 1948, among others).

The tavern was not merely a place where people gathered to share a drink after work or in the evenings. As in many working class communities, the tavern was a multifunctional institution that stood at the hub of daily social life in the coalfields (Cumbler 1979). Among other things, taverns stood as places where men could gather to exchange news and information about the workplace; they served as labor exchanges, as banks and post-offices, as boarding houses, and as sources of information about housing for new immigrants to the coalfields. Taverns often sponsored sport teams, and tavern keepers sometimes functioned as scribes for the illiterate. Because taverns frequently stood at the center of local ward politics, tavern keepers could help miners make contact with local politicians and the local police, a service that was always helpful when a miner ran afoul of the law in one way or another.

Like taverns, immigrant churches were multifunctional institutions. Built not only as places of worship, immigrant churches, which often were constructed after the taverns, were lay initiatives, especially among Protestants and eastern, central, and southern Europeans (Greene 1968). Church congregations, like taverns, provided an opportunity to build ties among church organizers and parishioners, and created opportunities for building networks of social participation that would strengthen bonds of trust within the local congregation or parish. Churches of all creeds provided a variety of social services for their members at a time when the state and corporations offered little or none. Bonds were further strengthened in those congregations, almost entirely Catholic, that established elementary and secondary schools. The churches helped to found and sustain immigrant middle classes. Through their contracts for goods and services, they contributed to building and sustaining middle class entrepreneurs within their communities.

Another type of institution that promoted internal group solidarity among immigrant populations was the mutual benefit society, along with fraternal organizations and women's organizations. Scott Cummings suggests that mutual benefit societies were a direct product of the nineteenth-century laissez-faire economy that predominated in the United States (Cummings 1980).

Mutual benefit societies were far more than insurance funds to cover burial expenses or centers of social activities. They also were a means of communal self-discipline. Because most mutual benefit societies had rules providing for the suspension or expulsion of members for "improper" behavior, they functioned as self-policing institutions within ethnic communities (Stolarik 1980). Dismissal from a mutual benefit society for unruly behavior meant that a person lost all benefits and was forced to make do in times of need without the formal support of the community.

Mutual benefit societies were an important source of capital formation and capital circulation in immigrant communities. They played significant political functions within and for the immigrant communities, offering a political training ground and a source of votes for rising ethnic politicians, whose campaign materials always made prominent mention of their association with these organizations.

Ethnic identity and ethnic community formations have continued to persist in the five anthracite counties. In all of them there is substantial bundling of different ethnic groups. In 2000 in Carbon County, we find a significantly higher percentage of residents identifying themselves as Czech, English, Hungarian, Irish, Lithuanian, Polish, Russian, British (including the English, Scottish, Scots-Irish, and Welsh people), or Dutch (probably reflecting "Pennsylvania Dutch" ancestry rather than actual Dutch ancestry) and lower percentages of American and Slovak identity (U.S. Bureau of the Census 2000).

In Lackawanna County, a smaller percentage of residents identified their ancestry as Dutch (Probably Pennsylvania Germans), German, Lithuanian, or American. Higher percentages identified themselves as Irish, Italian, Polish, Russian, or British (ibid.).

Luzerne County had low percentages of self-identified Dutch (again, probably people whose ancestors were Palatine Germans, the so-called "Pennsylvania Dutch"), German, Ukrainian, and American ancestry groups. It has relatively higher percentages of people in the Irish, Italian, Polish, Russian, Slovak, and British ancestry groups (ibid.).

Northumberland County had a somewhat different ethnic profile than the other counties. It has a significantly lower percentage of Czechs, Irish, Italians, Lithuanians, Poles, Russians, Slovaks, and British than did other anthracite counties. On the other hand, it has significantly higher percentages of persons claiming German, Dutch, or American ancestry (ibid.).

Schuylkill County in 2000 had significant over-representations of Dutch, Germans, Lithuanians, and Ukrainians compared to the rest of the former anthracite counties. It had under-representations of Irish, Italians, Poles, Russians, and British stock (ibid.).

The patterns of ethnic concentrations remained relatively stable between 2000 and 2010 suggesting that the patterns of neighborhood community and ethnic segregation that existed earlier in the twentieth century have persisted to a high degree (U.S. Bureau of the Census 2000, 2010).

In sum, tavern, churches, ethnic schools, mutual benefit and benevolent societies, spatial segregation, and other formal and informal networks stitched together individuals, households, social strata, and class fractions into coherent communities that provided some degree of protection from the hardships of life in the anthracite country. Without even the minimal assistance received in and through these organizations, life could be very hard indeed for the anthracite miner and his family.

The ethnic communities were important not only for the assistance they provided their members and their families. They also could be fashioned into collective weapons in the ongoing struggles against the mining corporations (Greene 1968; Keil 1982; Miller and Sharpless 1985). In their struggles with the mining companies, these communities often blended non- or preindustrial traditions and practices with more modern strategies and tactics, as we previously noted. Ethnic groups, except in Schuylkill and Northumberland Counties, were forged into an alliance with the Democratic Party and unions, which contributed to the undoing of the hold the bourgeoisie and its class allies had on politics in the coalfields.

Despite the eclipse of bourgeois direct political rule in much of the region, the bourgeoisie did not lose all power in the area. In Luzerne County, for example, the power and influence of the local bourgeoisie was demonstrated by the events taking place in the aftermath of the floods caused by Hurricane Agnes in the summer of 1972. The bourgeoisie seized control of the redevelopment agenda; excluded major local political actors, rank-and-file community members, and trade union leaders; and reshaped the development of downtown Wilkes-Barre to suit their own interests, mainly through an attempt at the preservation of downtown property values and those in other areas where they had significant investments. Their efforts to revitalize the central city of Wilkes-Barre ultimately were a failure. Rental values are low, there is a popular perception that the downtown is unsafe, the architectural value of buildings put up when coal was in its heyday has been seriously compromised, historic buildings have been torn down, and the city's central business district is populated by retail businesses of a marginal character, with the exception of the Kirby theater, a movie complex, one department store, several office buildings, and a bookstore. All of this reflects the continuing decline of Wilkes-Barre as a retail center for goods and services.

Figure 6.1 **Luzerne County Courthouse, built in 1899, reflecting the county's prosperity.**

Figure 6.2 Three preteen workers in a mining patch from the Ewen Breaker in Pitttson, Pennsylvania.

Figure 6.3 Young boys working in a dust filled room in a breaker while supervised by men with clubs to prod them to keep up the pace.

Figure 6.4 Group of breaker boys.

Figure 6.5 The Huber Colliery in more prosperous times in the coal industry.

Figure 6.6 International Ladies' Garment Workers rallying during a strike.

Figure 6.7 Centralia. Area of a former town dismantled because of a fifty-year underground mine fire in Columbia County.

Figure 6.8 Concrete City. Built in 1911 for "higher level" mine workers—foremen, technicians, and the like—by the DL and W Railroad. Housing was restricted to native English speakers. Located near Hazleton, Pennsylvania.

Figure 6.9 Radisson Lackawanna Station Hotel built in 1908 by the DL and W Railroad. Used as a train station and corporate offices until 1970.

Chapter 7

Subsequent Waves of Deindustrialization

The former anthracite region of Pennsylvania underwent a second wave of deindustrialization. The first took place with the decline of the coal industry, which began after World War I and continued into the 1950s and 1960s. According to researchers at the U.S. Geological Survey (1997), "The historical production curve for Pennsylvania anthracite is a classic example of a nearly normal discovery to depletion life-cycle. The coal resource was developed in the middle of the 19th Century and went through a 70-year long period of rapid growth until 1917, when annual production reached 99.7 million tons during World War I." Production then dropped dramatically during the depression, only to recover briefly during World War II, when the industry once again began its long-term decline. There continue to be large deposits of anthracite left in the ground, but, for a variety of reasons, including the geology and terrain of the area and the depletion of more accessible veins of coal, it is not recoverable at costs that will yield a profit.

During this long period of decline, there was a great deal of labor instability in the coal industry, especially in the 1920s and 1930s, as capital intensified its exploitation of the mine workforce. Capital, intent on maintaining profits, degraded the labor process in mining and struggled to continually reduce wages, absolutely and relatively. The workers resisted by fighting back against their employers and against their union, which many miners, for good reason, believed was collaborating with the coal operators. In the eyes of many mineworkers, the UMWA had a long history of cross-class collaboration with major coal operators.

John Mitchell, who led the strikes of 1900 and 1902 and who went on to become a long-time president of the UMWA was a strong proponent of conservative, bureaucratic, "responsible" business unionism. Mitchell, along with Samuel Gompers, the president of the AFL, was an early member of the National Civic Federation (NCF). The NCF was an organization comprised of major industrialists, financiers, politicians, and selected labor leaders. The NCF was committed to fighting socialism and, later, Bolshevism, and to creating conditions that would permit the stabilization and expansion of newly emerging corporate capitalism.

The leading spokesman for the NCF was Ralph M. Easley, a highly conservative Republican. A number of prominent Americans belonged to the NCF, including the coal baron senator Marcus A. Hanna, a Republican from Ohio; George M. Perkins, a partner of J. P. Morgan; and former Presidents Grover Cleveland and William H. Taft.

There were also representatives of organized labor in the NCF, such as Samuel Gompers and Mitchell. The NCF's position was that it was not opposed to organized labor, per se, but only to what it saw as "irresponsible unionism." According to Hanna, "responsible unionism" was unionism in which union members were expected to support the demands of union leaders, who acted as the "labor lieutenants of the captains of industry" (Thompson and Murfin 1976, 11).

In effect, such unionism was a complaint unionism in which unions deferred to the demands and needs of capital and in which the "real needs" of workers were identified by the so-called captains of industry rather than by the working class itself. Under such a definition of responsible unionism, the goals of unions should include remaining a part of the newly emerging corporate processes of production, rather than fundamentally challenging these processes and the subordinate place of workers in them.

While Mitchell was the only UMWA president to be a member of the NCF, he was not the only conservative elected to lead the UMWA. A number of the mineworkers' union presidents were conservative Republicans and corporate sympathizers. In 1911, Mitchell was forced to resign from the NCF by the rank-and-file membership or lose his union card. Even after resigning, however, his sympathies and conservative politics did not change. He continued to believe that only in collaboration with the state and capital could labor truly advance its interests.

Another conservative that held the UMWA presidency was Thomas Lewis of Locust Gap, Pennsylvania. He was one of the early founding members of the UMWA in 1890 and served as president of the union from 1908 until 1911. He lost the union presidency after signing labor agreements in two regions of the coalfields without the support of the regional presidents. He eventually quit the union and became an antiunion consultant for coal producers in West Virginia, published a trade journal for coal operators called the *Coal Mining Review*, and helped to found the antiunion National Coal Association.

Frank Hayes was an exception to the pattern of conservative leaders of the UMWA. He ran for governor of Illinois on the socialist ticket in 1919 while president of the UMWA. During his campaign, he turned over the presidency to John L. Lewis, who was elected president of the UMWA in his own right in 1920, just around the time that the anthracite industry began its long-term decline. Lewis's tenure was anything but peaceful. He presided over an unruly workforce, many of whom saw him as a collaborator with the coal companies. His goal, like that of Mitchell, was to ensure inclusion of the union in the production process but not to overturn that process. He cooperated with the operators in helping to drive the UAMP and the National Miners' Union out of the anthracite fields, and dismantled whatever was left of the union's dissident movements. He was a notorious red-baiter. Lewis served as president until 1960, presiding over the demise of anthracite production in northeastern Pennsylvania and leaving behind an undemocratic, corruption-ridden union in the anthracite fields.

Large profits were made possible by the continuing exploitation of the working class in the mining industry, by intensification of the mining labor process, and by the development of monopoly production in the industry. Before World War I, the railroads active in the coalfields had acquired control over 90 percent of un-mined anthracite coal reserves, were producing around three-quarters of the annual output of anthracite, and controlled more than one half of the remainder through long-term sales and carrying contracts (Rochester 1931, 69).

Furthermore, since 1920 the railroad companies and the mining companies had been forced to separate themselves and operate as independent entities. But this separation was more theoretical than actual, as was the separation of the mining companies and the railroad companies from each other. Morgan banks and interests continued to control the railroads, and, directly or indirectly, they still controlled the mining companies as well. Moreover, the big coal producers functioned as a price-fixing cartel. They were integrated through cross-membership on their boards-of-directors and through the cross-memberships they stitched together in the anthracite industry (ibid., 69).

By the mid- to late 1950s, little was left of the anthracite industry. Most of the major production companies had shut down their operations or were leasing their coal rights to tenants, many of whom were undercapitalized. These tenants degraded working conditions even further and many engaged in blatantly unsafe mining practices, often in collusion with corrupt local union officials. Such was the case at the Knox Mine. The investigations that followed this disaster revealed an industry that seemingly was able to survive only by gutting union contracts with sweetheart deals with tenant operators and an industry and a union that was allegedly deeply corrupted by organized crime.

The coal industry left a profound mark on the former anthracite region. It degraded the landscape and left an ecological and economic disaster area after the major operators cleared out of the region. The greatest positive effect of the mining era was the creation of conditions that led to the formation of a militant union culture in the region. This culture, including a carefully crafted collective memory fashioned by the *Citizens' Voice,* a union-owned strike paper born in a conflict between the media giant Capital Cities and its Wilkes-Barre newspaper unions (Keil 1988), led to a community that could be easily transformed into a platform of collective action and mobilized to defend workers when it became evident that a corporation or group of corporations were transgressing the moral rules that the community applied in assessing labor-management relations. Time and again, the community could be mobilized to support workers on strike.

A major strike occurred in the region beginning in 1978. Capital Cities, a large media conglomerate, purchased Wilkes-Barre's *Times-Leader* in 1978 for an estimated $10 million, a substantial discount, estimated to be anywhere from $10 to $20 million. A monopoly paper of this size should have gone for somewhere around $30 million (Keil 1988).

Shortly after the purchase, the new owners began contract negotiations with the paper's four unions, demanding significant concessions, especially from the Newspaper Guild's Local 120, which represented writers, photographers, and white collar workers, including the clerical staff. This local had a well-deserved national reputation for effective labor militancy and pan-union activism. Its contracts consistently were among the strongest for a paper of its size in the country. Capital Cities was convinced that without concessions from the unions, it would not be able to maximize profits in Wilkes-Barre and northeastern Pennsylvania.

Capital Cities was convinced when it arrived in Wilkes-Barre that it could easily tame and even possibly eliminate the unions at the *Times-Leader.* After all, across the country, newspaper owners had unions on the ropes, getting major contract concessions or breaking them altogether. Capital Cities, itself, had subordinated its printers in Fort Worth, Texas, and had defeated the Newspaper Guild and the International Pressmen and Assistants Union in Pontiac, Michigan, a union stronghold. So it saw no reason why it could not repeat its successes in Wilkes-Barre.

The newspaper unions maintained their solidarity and were able to capitalize on public memory to draw parallels between Capital Cities and the old coal companies, much to the newspaper unions' advantage. Capital Cities was framed as a hyper-exploiter of local labor and, equally importantly, a violator of the local moral economy, which contained a set of expectations as to how a company ought to treat its workers. When the workers struck and set up their own newspaper, the *Citizens' Voice*, they picked up support from many of the local communities in and around Wilkes-Barre. Even the adolescent newspaper carriers defected en masse to the *Citizens' Voice* and many brought their customers with them to the new paper (Keil 1984). No longer a worker paper, it was sold to a newspaper publishing company. *The Citizens' Voice* has managed to hold on to the present day (see Keil 1984, 1988), although it is no longer union-owned. According to the Alliance for Audited Media (September 30, 2012), the former worker paper has an average Monday through Saturday circulation of 42,254, while the former Capital Cities' paper has a Monday through Friday circulation of 38,838 and a Saturday average circulation of 30,757. *The Citizens' Voice* Sunday edition lags behind the *Times-Leader*, with the former having a circulation of 43,432 papers and the latter having a circulation of 55,488 papers (September 30, 2012).

To the local bourgeoisie, the culture of unionism has been a constant source of irritation. They place much of the blame for the inability to redevelop the region on its reputation for labor militancy and community support for such militancy. As employment in the anthracite industry declined, local economic development groups launched community drives to raise funds to entice new employers into the region. Few local capitalists were willing to invest substantial amounts of their own money in trying to rehabilitate the local economy. Economic development groups claimed that one of the reasons for their inability to bring in "good" jobs, besides the regions' reputation for labor militancy, was the low skill levels of the local population of workers, especially male workers. It was said by prospective investors that few of the former mineworkers had skills easily transferable to factory work, and many lacked the work habits necessary for modern factory employment. Given such a reputation, the major industries that the bourgeoisie recruited were lower-wage industries, especially the garment industry, which hired women at wages below the rates paid in New York and Philadelphia, two metropolitan areas from which the garment industry fled in order to avoid unions, their higher wage rates, and demands for good working conditions.

It, however, was not long before the women in the region began organizing with the assistance of Min Mattheson from the International Ladies Garment Workers' Union (ILGWU). They engaged in a bitter strike to gain union recognition and increased wages for the garment industry in the former coalfields. As often was the case in the heyday of mining, the epicenter of the militance was in Pittston. There the strike not only confronted the power of regular operators but also the power of a strong organized Mafia family, with heavy investments in garment manufacturing, which had organized its own company union to try to counteract the influence of the ILGWU. The ILGWU prevailed in the strike and went on to become a powerful economic and political force inside the Democratic Party, until disinvestment led to the dismantling of the industry in the regions. The decline of the garment trades represented the second wave of deindustrialization to sweep through the former coalfields.

The local bourgeoisie also tried to recruit other industries, especially those that would absorb the excess labor power of men. But, rather than investing large parts of their own capital to start businesses, the local bourgeoisie worked to recruit outside investors to whom they offered heavy government subsidies, as well as subsidies raised from the public through industrial development fund-raising drives. Using public subsidies, local development groups opened industrial parks, some of which were more successful than others, and attracted several major corporations, few of which have survived to the present day.

The continuing economic decline of the region led to the export of a large part of its population. Except for Carbon County, which has benefited from population growth in the Pocono Mountains, the four other counties of the anthracite region have had stagnant or declining population numbers. As is usually the case in declining areas, the larger part of the out-migration stream has been made up of younger people. This has put pressure on significant parts of the local social fabric, as churches have shut down or been consolidated and public and parochial schools have closed or consolidated. With an aging population, resources, both economic and human, for investment in local institutions have become scarcer over time.

Many of the manufacturing companies that were recruited to replace the garment industry succumbed to a third wave of deindustrialization. Between 1997 and 2007, Lackawanna and Luzerne Counties each experienced a drop of 26.3 percent in manufacturing jobs. Carbon County manufacturing jobs decreased by 32.8 percent, while Schuylkill County manufacturing jobs decreased by 28.9 percent and Northumberland by 22.6 percent (U.S. Bureau of the Census 1997, 2007). By 2012, manufacturing accounted for 11.8 percent of the workforce in Lackawanna County, 13.5 percent in Luzerne County, 19 percent in Schuylkill, 16.5 percent in Northumberland, and 13.1 percent in Carbon (U.S. Bureau of the Census 2012).

The former anthracite region has responded by selectively shifting to service jobs that are largely nonunion, pay low wages, have little opportunity for upward mobility inside of the enterprise, and offer few fringe benefits or a stable working environment. Between 1997 and 2007, Carbon County increased 9.5 percent in retail sectors employment and Lackawanna County increased by 7.8 percent. Luzerne County was relatively stable with a 0.9 percent decrease in retail sector jobs, partly reflecting the decline of Wilkes-Barre as a retail center. Northumberland and Schuylkill Counties both had shrinking retail sector employment, with drops of 10.9 percent and 14.6 percent respectively so that even these typically low-wage jobs were declining in these counties (U.S. Bureau of the Census 1997, 2007). By 2012, retail employment was 13.8 percent of the workforce in Schuylkill County, 12.9 percent in Northumberland, 13.4 percent in Luzerne County, 12.2 percent in Lackawanna, and 13.1 percent in Carbon County. These figures compare to 11.6 percent in Pennsylvania, as a whole (U.S. Bureau of the Census 2012).

Employment in the former anthracite region grew strongly in the health care and social assistance sectors. Carbon County's employment in this area, between 1997 and 2007, grew by 310 percent, Lackawanna County's growth was 250 percent, Luzerne County's was 188 percent, Northumberland County's was 130 percent and Schuylkill County's was 158 percent (U.S. Bureau of the Census 1997, 2007). The 2012 *Quick Facts* shows that educational services, health care, and social assistance sectors are the modal categories of employment in all of the anthracite counties. These sectors comprises 23.9 percent of the workforce in Carbon County, 26.5 percent in Northumberland County, 23.6 percent in Schuylkill County, 24.2 percent in Luzerne County, and 27.8 percent in Lackawanna County, compared to 25.2 percent in Pennsylvania (U.S. Bureau of the Census 2012).

As a result of the decline of manufacturing and the growth of lower-wage service sectors, the former anthracite coal regions continue to lag behind the rest of Pennsylvania in almost all economic indicators. For example, in the first quarter of 2008, the average weekly wage in Pennsylvania was $869, while in the former anthracite counties it was $621 (U.S Bureau of Labor Statistics 2008). Luzerne and Lackawanna Counties, with wages of $674 and $645 respectively, ranked at the bottom two of Pennsylvania's nineteen largest counties, and were in the bottom third of counties nationwide. The other three coal-region counties had even lower weekly wage levels. Carbon's was $550, Northumberland's was $622, and Schuylkill's was $617.

In 2012, median family income in Pennsylvania was estimated at $65,105, while in Carbon County it was $57,280, in Northumberland County it was $54,499, in Schuylkill it was $55,824, in Luzerne County it was $56,493, and in Lackawanna it was $59,081. With the exception of Carbon County, the northern field counties are doing better than the southern/central ones. Family income for the entire region, however, lags behind the Pennsylvania average. For median earnings per worker, the Pennsylvania figure is $30,292. In Lackawanna County it is $27,301, in Luzerne County $27,507, in Schuylkill County $28,131, in Northumberland County $25,712, and in Carbon County $28,600 (U.S. Bureau of the Census 2012).

The former anthracite counties showed significant differences in the proportions of their workforce with bachelor's degrees or higher in 2012. In Lackawanna County, 24.6 percent of its population age twenty-five and above had post-secondary degrees. Luzerne County had 20.2 percent, Carbon County 15 percent, Northumberland County 14 percent, and Schuylkill County 14.3 percent. The commonwealth is 26.7 percent (ibid.).

The data show that the southern/central field counties lag behind the northern counties by substantial amounts in terms of percent of the population with post-secondary educations.

The counties also differed in unemployment in 2010. While commonwealth-wide average unemployment stood at 9.4 percent in March 2010, Carbon County's was 12.5 percent, Northumberland's was 11.0 percent, Schuylkill's was 11.5 percent, Lackawanna's was 9.7 percent, and Luzerne's was 11.0 percent (U.S. Bureau of Labor Statistics 2010).

Reflecting differences in "initial conditions" and trajectories of development, including patterns of deindustrialization and local responses to it, there are additional marked variations in the levels of development of Luzerne and Lackawanna Counties versus the remainder of the former anthracite region, with Luzerne and Lackawanna Counties generally performing better economically than the other parts of the area. According to analysis of the 2000 U.S. Census data from the communities in Lackawanna and Luzerne Counties, the northern region ranked significantly better than the southern/central region on a number of indicators: labor force participation, median household income, median family income, median per capita income, median value of owner-occupied housing, female labor force participation, and female median income. These differences largely persist into the present.

Median income for full-time male workers was $48,283 in Pennsylvania in 2012. In Carbon County it was $43,445, in Northumberland County $39,419, and in Schuylkill County $43,040. In Luzerne County it was $42,573 and in Lackawanna County $43,840 (U.S. Bureau of the Census 2012). Variations in median income for full-time male workers are not related to area, but all counties fall below the commonwealth average.

For full-time female workers, median income was $36,858 in Pennsylvania in 2012, yielding a male-to-female income ratio of 1.31 to 1. In Lackawanna County, the number was $32,927 for female workers, for a male-to-female income ratio of 1.33 to 1. This exceeds the commonwealth average. In Luzerne County, female median income was $32,331 in 2012, for a ratio of 1.32 to 1, which approximated the Pennsylvania ratio of 1.31 to 1. In Carbon County, median income for full-time women workers was $32,416 and the ratio 1.34 to 1. Women full-time workers' median income was $29,993, for a male-to-female ratio of 1.31 to 1 in Northumberland County. Schuylkill County's full-time working-women's median income is $31,447, and the ratio there was the worst at 1.37 to 1 (ibid).

At the county level, comparing 2000, 2008, and 2012, little changed in the indicators for which data are available over the time period. For the most part, the northern region continued to outperform the central and southern regions. On virtually all of the indicators, the former coal regions lagged behind state-wide averages (U.S. Bureau of the Census 2000, 2010).

The northern fields in 2000 had higher percentages of workers in "management, professional, and related occupations," with 29.4 percent of the northern regions' workers falling into this category and 22.13 percent from the central and southern regions. Northern field workers also were more likely than central and southern field workers to have "sales and office occupations." The respective percentages in 2000 were 26.7 percent and 22.4 percent. Central and southern field workers were more likely to be found in "construction, extraction, and maintenance occupations" than were northern workers. In the central/southern regions, 12.2 percent worked in such occupations, while in the northern fields, 10.5 percent worked in these occupations. Finally, in terms of occupational classifications, central and southern field workers were much more likely than their northern counterparts to be employed in "production, transportation, and materials moving occupations." There was no difference between the regions in percent of workers in "service" occupations (U.S. Bureau of the Census 2000).

These patterns held up in 2008. Looking at the county level, in Carbon County the proportion of employees in "management, professional and related occupations" in 2000 was 22.8 percent and increased to 24.6 percent in 2008. In Lackawanna County in 2000, 29.6 percent of workers were in this category, which, increased slightly to 30.6 percent in 2008. Luzerne County's workforce contained 27.7 percent in this category in 2000 and 28.5 percent in 2008. Northumberland County had 22.9 percent of its workforce in these occupations in 2000 and 24.6 percent in 2008. Finally, in Schuylkill County in 2000, 22.9 percent of workers were in these occupations, while in 2008, there was 27.4 percent. Schuylkill County showed the highest rate of growth of the five counties, followed by Carbon County (American Fact Finder 2010).

In 2012, in Schuylkill County 26.6 percent of the workforce fell into the category of "management, business, and arts occupations." Northumberland County had 30 percent of its workforce in this category of employment, while Carbon County had 27.4 percent. Luzerne County had 30 percent and Lackawanna County had 35 percent (American Community Survey 2012). The two former northern field counties were doing better at generating jobs such as these than were the two of the three southern and central counties.

With respect to "service" occupations, 14.9 percent of Carbon County's workforce was employed in this category in 2000. In 2008, the corresponding figure was 15.3 percent, a slight change over the time period. In 2000, 15.6 percent of Lackawanna County's workforce was in "service" occupations; by 2008, this had increased to 17.5 percent, a change of 1.9 percent. Luzerne County was close to Lackawanna County with 15.2 percent of its workforce in "service" occupations in 2000. The same was the case in 2008, when 17.6 percent of Luzerne County's employees were in "service" occupations. In 2000, Northumberland County had a relatively high 17.1 percent of its workforce in "service" occupations. By 2008, this figure reached 18.7 percent, again, the highest percentage of any county in the former anthracite region for that year. In 2000, Schuylkill County had the lowest percentage of the workforce in "service" occupations, 14.4 percent. The corresponding figure in 2008 was 15.2 percent (U.S. Bureau of the Census 2000; American Fact Finder 2010).

Lackawanna County had 19.8 percent of its 2012 workforce in "service" occupations. Luzerne County had 17.8 percent, Carbon County had 17.2 percent, Northumberland County was at 17.8 percent, and Schuylkill County at 18.1 percent (U.S. Bureau of the Census 2012).

"Sales and office" occupations made up 24.6 percent of employees in Carbon County in 2000 and 25 percent of employees in 2008. As in 2000, in 2008 it was one of the largest categories of workers in the county. Lackawanna County's workforce in 2000 was comprised of 29.1 percent "sales and office" employees, a figure substantially higher than that of Carbon County. In 2008, the percentage was 29 percent, virtually unchanged from 2000. Luzerne County, in 2000, had 28.1 percent of its workers in this occupational category; in 2008, the percentage was basically unchanged at 28.2 percent. "Sales and office" employees comprised 22.8 percent of the workers in Northumberland County in 2000 and 23 percent in 2008. Once again, there was little change between the two time points. In Schuylkill County, 22.8 percent of workers were "sales and office" employees in 2000. The same percentage was observed for 2008. Overall, the percentages of employees in this category were highly stable over time across counties. No county exhibited marked changes between 2000 and 2008 (U.S. Bureau of the Census 2000; American Fact Finder 2010).

In 2012, in Schuylkill County, 20.4 percent of workers filled "sales and office occupations," a figure lower than 2008 but one that held the third largest number of workers in the county. In Northumberland County, the percent of workers in this category was 26.6, an increase from 2008. Carbon County had 24.7 percent of workers in "sales and office occupations," an increase since 2008. Luzerne County had 26.6 percent of its workers in this category, a figure virtually unchanged from 2008. Finally, Lackawanna County had 24.1 percent, for a decrease from 2008 (U.S. Bureau of the Census 2012).

"Natural resources, construction, and maintenance" occupations made up 12.9 percent of all workers in Carbon County in 2000. By 2008, this figure had increased to 15.4 percent, a gain of 2.5 percent. This increase reflected the growth of Pocono Mountain communities as vacation areas and homes to a large number of commuters to New York and northern New Jersey. These workers were only 8.2 percent of all employees in Lackawanna County in

2000 and 2008. In Luzerne County in 2000, they were 9.8 percent of the workforce, less than Carbon County, but more than Lackawanna. By 2008, the percentage dropped to 8.9 percent in Luzerne County, bringing it closer to Lackawanna's numbers. Northumberland County's percentage of workers in these occupations was 9.9 percent in 2000, approximately equal to Luzerne County in that year. In 2008, it was virtually unchanged at 9.8 percent. In Schuylkill County, 11.7 percent of workers were in these occupations in 2000. By 2008, this rate was almost unchanged at 11.6 percent (U.S. Bureau of the Census 2000; American Fact Finder 2010).

In 2012, the occupational category "natural resources, construction, and maintenance occupations" made up 7.9 percent of Lackawanna County's workforce and 7.3 percent of Luzerne County's. Such workers made up 13.2 percent of Carbon County employees, 7.3 percent of Northumberland County's employees, and 11.2 percent of Schuylkill County's employees (American Community Survey 2012).

"Production, transportation, and material moving" occupations were 24.5 percent of Carbon County's workforce in 2000. By 2008, this figure had decreased to 19.4 percent, a drop of 5.1 percent. This was the largest decline of any of the counties. Workers in these categories were 17.4 percent of all employees in Lackawanna County in 2000 and 14.5 percent of all employees in 2008, a decline of 2.9 percent. In 2000, these workers were 19 percent of employees in Luzerne County, more than Lackawanna but less than Carbon County. Such employees were 16.6 percent of the workforce in 2008, or 2.4 percent less than in 2000. They were 26.5 percent of all employees in Northumberland County, making them the largest group in the workforce in 2000. In 2008, they were 23.5 percent, 3 percent less than in 2000. They had fallen behind "management, professional, and related" occupations as the largest category of workers. Schuylkill County, in 2000, had the highest proportion of workers of any county falling into this category at 26.9 percent. By 2008, there was a major drop to 22.2 percent. This decline 4.7 percent was second only to the decline in Carbon County (U.S. Bureau of the Census 2000; American Fact Finder 2010).

"Production, transportation, and material moving occupations" employed 23.7 percent of Schuylkill County workers in 2012. Northumberland County employed 18.3 percent of its workforce in this category in 2012. Workers employed in these occupations in Carbon County in 2012 made up 17.5 percent of the workforce. Luzerne County had 18.3 percent of its workforce in these occupations in 2012. This was an increase from 2008. Lackawanna County had 13.2 percent of its workforce in this category in 2012 (American Community Survey 2012).

Table 7.1, which follows, presents data on the Location Quotients (LQs) for the Scranton/Wilkes-Barre Metropolitan Statistical Area (MSA) and

Table 7.1 Location Quotients

Occupations	*LQ MSA*	*LQ Pennsylvania*
Management	0.623	0.786
Business and Finance	0.697	0.941
Architecture and Engineering	0.679	0.948
Life, Physical, and Social Science	0.383	0.902
Community and Social Service	1.527	1.348
Legal	0.618	0.875
Education, Training, and Library	0.891	0.949
Arts, Design, Entertainment, Sports, and Media	0.726	0.828
Health Care Practitioners and Technical	1.134	1.105
Health Care Support	1.181	1.207
Protective Service	0.897	0.866
Food Service and Serving	0.916	0.970
Building and Grounds Cleaning and Maintenance	0.894	0.972
Personal Care and Service	1.053	1.032
Sales and Related	1.069	1.003
Office and Administrative Support	1.068	1.030
Construction and Extraction	0.838	0.948
Installation and Maintenance	1.061	1.050
Production	1.304	1.092
Transportation and Materials Moving	1.407	1.065

Source: Bureau of Labor Statistics, May 2010.

the Commonwealth of Pennsylvania. These LQs are a measure of the relative concentration of occupations in a geographical area compared to the United States.

The following occupations are more concentrated in Pennsylvania than in the United States: community and social service; health care practitioners and technical; health care and support; personal care and service; office and administrative support; installation, maintenance, and repair; production; and transportation and materials moving occupations.

In the Scranton and Wilkes-Barre MSA, the following occupations have the highest concentrations compared to the United States: community and social service; health care practitioners and technical; health care and support; personal care and service; sales and related; office and administrative support; installation, maintenance, and repair; production; and transportation and materials moving occupations. The only difference in the two lists is for sales, where there is an over-representation in the MSA, but not for the commonwealth.

Comparing LQs in the MSA to those in the commonwealth, we see that the higher status occupations of management; business and finance; architecture and engineering; life, physical, and social science; legal; education, training, and library; and arts, design, entertainment, sports, and media are more concentrated in the commonwealth than in the MSA. Only for health practitioners and related is this not the case. The LQ is higher in the MSA than the commonwealth for this set of occupations.

In the case of other occupations such as protective service; personal care and service; sales and related; office and administrative support; installation and maintenance; production; and transportation and materials moving there are higher concentrations in the MSA than in the commonwealth.

In all occupational categories, the MSA mean annual incomes are lower than those in the commonwealth as a whole. The greatest differences generally are in the higher status occupations such as legal; management; business and finance; life, physical, and social science; arts, design, entertainment, sports, and media; as well as in protective services; sales and related; and

Table 7.2 Mean Annual Incomes for Scranton/Wilkes-Barre MSA and Pennsylvania

	Mean Annual Income	*Mean Annual Income*	
Occupation	*Scranton-Wilkes-Barre*	*Pennsylvania*	*Percent Difference*
All Occupations	$36,880	$43,050	86
Management	$82,870	$104,110	80
Business and Finance	$53,860	$66,120	81
Architecture and Engineering	$66,250	$69,610	95
Life, Physical, and Social Science	$54,620	$66,610	82
Community and Social Service	$35,270	$39,250	90
Legal	$75,250	$99,860	75
Education, Training, and Library	$48,570	$53,520	91
Arts, Design, Entertainment, Sports, and Media	$37,160	$45,830	81
Health Care Practitioners and Technical	$61,710	$67,380	92
Health Care Support	$26,250	$26,820	98
Protective Service	$36,460	$42,110	87
Food Service and Serving	$20,240	$21,660	93
Building and Grounds Cleaning and Maintenance	$24,090	$26,090	92
Personal Care and Service	$23,410	$23,910	98
Sales and Related	$31,440	$37,250	84
Office and Administrative Support	$30,060	$33,220	90
Construction and Extraction	$39,340	$44,430	89
Installation and Maintenance	$38,020	$41,550	92
Production	$32,610	$34,830	94
Transportation and Materials Moving	$30,180	$32,720	92

Source: Bureau of Labor Statistics (May 2010).

construction and extraction. The other occupations come within 10 percent of the commonwealth's mean annual wage.

Table 7.3 presents the ratio of occupational incomes to the average annual income within the MSA and in the commonwealth. Examining Table 7.3 we see that the higher status occupations, except for architecture and engineering and health care practitioners and technical occupations, all have higher multiples of the mean income in the commonwealth than is the case in the MSA. At the same time, many of the lower status occupations come closer in the MSA to the average income than we find in Pennsylvania as a whole. These patterns of lower multiples for higher standing occupations and higher multiples for lower standing occupations suggest there is more equality in the MSA than there is in the commonwealth.

As the case with the Great Depression, when the anthracite region experienced a declining economy before it set in nationally, the Great Recession had a similar impact on the former coal regions, which began experiencing significant economic decline before the "official" onset of the Great Recession. As unemployment increased, incomes declined in real dollars in all

Table 7.3 Income to Average Income for the MSA and Pennsylvania

Occupations	*Scranton-Wilkes-Barre Income Ratios*	*Pennsylvania Income Ratios*
Management	2.25	2.43
Business and Finance	1.46	1.54
Architecture and Engineering	1.80	1.62
Life, Physical, and Social Science	1.48	1.55
Community and Social Service	0.96	0.91
Legal	2.04	2.32
Education, Training, and Library	1.32	1.24
Arts, Design, Entertainment, Sports, and Media	1.01	1.06
Health Care Practitioners and Technical	1.67	1.57
Health Care Support	0.71	0.62
Protective Service	0.99	0.98
Food Service and Serving	0.55	0.50
Building and Grounds Cleaning and Maintenance	0.65	0.61
Personal Care and Service	0.63	0.56
Sales and Related	0.85	0.87
Office and Administrative Support	0.82	0.71
Construction and Extraction	1.07	1.03
Installation and Maintenance Repair	1.03	0.97
Production	0.88	0.81
Transportation and Materials Moving	0.85	0.70

Source: Bureau of Labor Statistics (May 2010).

but Carbon County (measured in constant 1997 dollars), between 1998 and 2008, and work shifted to lower-paying economic sectors, bank deposits in the five counties shrank in the former anthracite region. While bank deposits increased in Pennsylvania by 19 percent from 1998 to 2008, they declined in Carbon County by 6.8 percent, in Lackawanna County by 3.2 percent, in Luzerne County by 2.1 percent, in Northumberland County by 13.0 percent, and in Schuylkill County by 16.3 percent (County Profiles 2009, 5, 10). These changes affected local credit and possibilities for local capital formation and investment, thereby shrinking the chances of locally stimulated growth and making the region even more dependent on extra-local capital. This situation already accelerated with the takeover of local banks by national banking conglomerates in the last quarter of the twentieth century. Such takeovers weakened the local bourgeoisie economically and socially. Banking had been an industry controlled by the local bourgeois, and the industry worked to economically and socially integrate the old bourgeoisie by cross-ownership and cross-directorships. They served to integrate elites into the Wilkes-Barre bourgeoisie across Luzerne County, especially when the Wilkes-Barre banks began consolidating smaller community banks under their control. In the modern era, local banks have fallen under the hegemony of regional and national capital.

As economic conditions deteriorated, poverty rates increased across the former anthracite region between 1997 and 2007. While Pennsylvania's poverty rate changed by 0.7 percent over this period, growing from 10.9 to 11.6 percent. In Carbon County poverty increased from 9.3 to 10.6 percent, for an increase of 1.3 percent. In Lackawanna County, poverty increased by 2.5 percent, from 10.1 to 12.6 percent. Luzerne County's poverty level increased from 10.8 to 11.7 percent, close to the level of change in the commonwealth. Northumberland County's poverty rate grew from 11.5 to 12.9 percent (the highest level in the five-county region). This was an increase of 1.4 percent. In Schuylkill County, poverty increased from 10.2 to 11.9 percent, a change of 1.7 percent, which was the largest percentage increase in the former anthracite fields (ibid., 5–6).

In 2012, the poverty rate for Pennsylvania was 13.2 percent. In Carbon County it stood at 13.1 percent. Schuylkill County's poverty rate for all people was 15.2 percent. Northumberland's was 16.9 percent. Lackawanna County's poverty rate for all people in 2012 was 14.7 percent and Luzerne County's was 16.4 percent, the second highest in the region (American Community Survey 2012).

With the dismantling of the industrial base across the former anthracite fields, relatively low wages, and growing poverty, welfare dependency increased in the five counties. In Pennsylvania, the change in enrollment in Children's Health Insurance Program (CHIP) from March 2008 to

March 2009 was 11.2 percent. In Carbon County, it was 11.3 percent, in Lackawanna County 7.2 percent, in Luzerne County 15.8 percent, in Northumberland County 30.7, percent, and in Schuylkill County15.6 percent (Center for Rural Pennsylvania 2009).

There was also an increase in dependency on food stamps with the onset of the Great Recession. Between June 2005 and June 2008, the percentage of families receiving food stamps in the commonwealth rose from 8.6 to 9.6 percent. In Carbon County, this percentage increased by 3.3 percent, from 8.0 to 11.3 percent. Lackawanna County's increase was from 8.0 to 10.3, percent, in Luzerne County from 9 to 11.3 percent, in Northumberland County from 5.7 to 7.6 percent, and in Schuylkill County from 6.9 to 9.3 percent (ibid., 6).

In order to reverse the effects of economic decline in the commonwealth, in the late 1990s, Governor Tom Ridge's administration developed a sequence of legislation to stimulate economic development and revitalize economically marginal communities throughout Pennsylvania. In October 1998, Governor Ridge signed Keystone Opportunity Zone (KOZ) legislation. In 2000, this legislation was amended with the passage of the Keystone Opportunity Expansion Zone (KOEZ) and Technical Change Bill (Act 119 of 2000), which expanded the initial program. The program was further expanded with the passage of Act 219 of 2002, which created the Keystone Opportunity Improvement Zone (KOIZ) that permitted a one-time opportunity for an expedited implementation of an economic development initiative under the terms of the KOZ legislation (Keystone Opportunity Zone 2003, 1).

The KOZ/KOEZ programs were developed to provide inducements for businesses to invest in communities that were struggling economically. The principal inducements were tax abatements. According to the report, *KOZ/KOEZ Four-Year Report*, the KOZ and KOEZ programs offer the following tax benefits to businesses and individuals:

> Businesses (with some limitations) may be entitled to state sales and use tax, corporate net income tax, capital stock and foreign franchise tax, real property tax, local net profits tax . . ., business privilege tax, mercantile license tax, bank share credit, job creation tax credit, realty use and occupancy tax, and local sales and use tax (Allegheny and Philadelphia Counties) incentives. Businesses relocating into a KOZ/KOEZ must either expand employment by 20 percent or must make capital investments of at least 10 percent of the gross revenues of the business of the previous year to take advantage of the incentives.
>
> Individuals who are residents of a KOZ/KOEZ also receive tax benefits including state and local tax exemptions for personal income earned as a KOZ/KOEZ resident; interest, rent, and dividends while living in the KOZ/KOEZ; net gains from the sale, exchange, or disposition of intangible personal property; and net gains or income derived through estates and trusts received as

> a resident of a KOZ/KOEZ. An individual must reside in a KOZ/KOEZ for 184 consecutive days during the taxable year in order to take advantage of the incentives. Both residents and non-residents of the KOZ/KOEZ may be entitled to exemptions for net income from the operation of a qualified business attributable to business activity conducted within a KOZ/KOEZ; net gains from the sale, exchange, or disposition of real or tangible personal property located in a KOZ/KOEZ; net gains or income from rents received from the rental of real or tangible personal property allocable to a KOZ/KOEZ; and real property taxes. (ibid, 2)

The KOZ/KOEZ legislation provided for substantial tax expenditures for businesses operating and persons living, within designated development zones. Over the four-year period evaluated, the legislation produced 13,613 jobs, both realized and projected, and preserved 7,962 jobs throughout the Commonwealth of Pennsylvania. These are hardly striking figures. The evaluation of the success of the program does not identify or attempt to identify how many of these jobs would have been created without the incentives, how many jobs left the zones, what the wage differentials were between the jobs that left and those brought in, how many jobs moved into the zones from non-zone areas, and what jobs actually were created. So, there really is no way to know if the claims that the KOZ/KOEZ legislation was or was not a success are true.

Data suggest that the KOZ/KOEZ program did not staunch the loss of jobs, especially manufacturing jobs, in Pennsylvania as a whole and in the former anthracite coal region in particular. In Pennsylvania, as in the United States, job losses have been especially severe in recent years in the high-wage manufacturing sector (McMillion 2008, 1).

In 2000, the former northern field had higher proportions of workers employed in managerial and professional occupations and in sales and office jobs, while the former central/southern fields had greater proportions of workers engaged in construction and production transportation, and material moving occupations, with both regions having equal proportions employed in service jobs. In terms of economic sectors, the northern fields exceeded the central/southern fields in trade; retail trade; information, finance, real estate, and rental and leasing; professional, scientific, management, administrative, and waste management; education, health, and social services; and arts, entertainment, recreation, accommodation, and food services. The two have been equal in construction, transportation, warehousing, and utilities and other services (except public administration). The central/southern region had a higher percentage of workers in manufacturing than did the northern region.

Between 2000 and 2007, even before the Great Recession, Pennsylvania lost 207,300 manufacturing jobs, which amounted to 24.3 percent of its manufacturing base. These jobs have been "replaced" by lower-wage employment in health care and social assistance, where wages are 30.6 percent lower than the average manufacturing wage. Likewise, jobs in state and local government pay 22.7 percent less than the average manufacturing wage, and food service and drinking places pay 74.2 percent less than manufacturing statewide. These three sectors were the fastest growing in Pennsylvania between 2000 and 2007. Overall, the growth rate in jobs for Pennsylvania was 1.7 percent, hardly indicating significant economic progress. The situation was even worse in the former anthracite coalfields, where job growth was an anemic 0.7 percent over the same period in the Scranton/Wilkes-Barre MSA, which has a more vital economy than most other parts of the former anthracite region (ibid., 1, 2).

Over the seven-year period in which Pennsylvania lost 207,300 manufacturing jobs, the Scranton/Wilkes-Barre MSA lost 12,400 manufacturing jobs, or 27.7 percent of its manufacturing base. As in Pennsylvania as a whole, these lost manufacturing jobs were replaced with lower-wage jobs in the service sector and in other relatively low paying sectors. Thus, the former anthracite coalfields, like Pennsylvania as a whole, and, perhaps, even to a greater extent, have spent the last several decades trading down in the job market (see Keil 1988, 36 for a discussion of this process in the 1980s in Luzerne County). They have surrendered higher-paying jobs for lower-paying ones. Even jobs that normally pay well in other places provide lower levels of compensation in the former coalfields than they do elsewhere, as we showed in Table 7.3.

For this reason, the average net worth per household in the former coal regions lagged behind that of the commonwealth average of $216,000. Carbon County came closest to the commonwealth average with a net worth per household of $198,000; next was Lackawanna County with an average net worth of $180,000, then Luzerne County at $175,000. Schuylkill and Northumberland Counties were substantially lower with averages of $152,000 and $140,000 respectively (Macke et al. 2008).

In 2007, manufacturing employment had dropped by 32.8 percent from 1997, but it was the highest paying sector in Carbon County. The total annual payroll in manufacturing in 2007 was $95,141,000. This was $46,444,268 less than it would have been if Carbon County had not lost 1,196 manufacturing jobs. Retail employment's average pay stood at $20,937 (53.9% of the average manufacturing payroll). Between 1997 and 2007, the ratio of manufacturing to retail employment dropped from 1.73 to 1 to 1.06 to 1 as a result of declining manufacturing employment combined with rising employment in the retail sector (U.S. Bureau of the Census 1997, 2007).

Health care and social assistance compared to manufacturing jobs dropped from a ratio of 1 to 5.9 in 1997 to 1 to 0.97 in 2007, showing that health care and social assistance enterprises employed slightly more people in 2007 than did the manufacturing sector. The average payroll in health care and social assistance in 2007 was $31,130, 80.2 percent of the average manufacturing payroll (ibid.).

The real estate, rental, and leasing category compared to manufacturing went from a ratio of 1 to 45 jobs in 1997 to 1 to 21 in 2007. The average pay in the real estate sector in 2007 was $31,795, 81.9 percent of the average manufacturing pay (ibid.).

Once again, there was relative growth in a lower-wage sector compared to what was happening in manufacturing, the highest-wage sector in the county. Accommodation and food services in Carbon County dropped from a ratio of 3 manufacturing jobs for every one job in this sector to 1.6 to 1. The average annual wage in this sector was $14,587 or 37.6 percent of the average manufacturing wage (ibid.).

Finally for Carbon County, manufacturing jobs to service jobs dropped from a ratio of 20 to 1 to 7 to 1. Again, this indicates an expansion of low-wage jobs and a contraction of higher-wage manufacturing jobs. The average payroll in "other services" was $18,815 or 48.5 percent of the average manufacturing payroll.

In Lackawanna County, the pattern is similar, but not as stark, as it had a more diverse workforce in 1997 than Carbon County and the contraction in manufacturing (26.3%) was not as proportionately severe. Had Lackawanna County been able to hold on to the 16,052 manufacturing jobs it had in 1997, instead of losing 4,218, however, the total annual manufacturing payroll would have been at $595,561,304 rather than $439,069,000, or $156,492,304 more in 2007 than what it turned out to be. The ratio of manufacturing jobs to retail jobs fell from 1.2 to 1 to 0.80 to 1, indicating the displacement of manufacturing as the category with the most workers in the county over the decade 1997 to 2007. Retail jobs averaged $19,858 or 53.5 percent of manufacturing pay, which paid on average $37,102. Retail sector growth is often a low-wage substitute for manufacturing employment (ibid.).

Another sector that expanded dramatically in comparison to manufacturing in Lackawanna County was health care and social assistance. In 1997, there were 3 manufacturing jobs for every 1 health care and social assistance position. By 2007, this ratio had fallen to 0.64 to 1. Wages in the health care and social assistance sector were 93.7 percent ($34,759) of the average manufacturing payroll (ibid.). So, even though there was a trading down of jobs in terms of pay in the sector shift of jobs from manufacturing to health care and social assistance, it was not as costly as the trade down in other sectors of Lackawanna County's economy.

The ratio of manufacturing jobs to real estate jobs also declined between 1997 and 2007, dropping from 21.5 to 1 to 13 to 1. Real estate employment paid an average wage of $24,386, or 65.7 percent of the pay of jobs in the manufacturing sector (ibid.).

In Lackawanna County, the ratio of manufacturing jobs both to accommodation and food and other services jobs both declined from 1997 to 2007. The respective figures were 2.2 to 1 to 1.4 to 1 for accommodation and food, and 7.7 to 1 to 5.2 to 1 for other services. The average pay for accommodation and food services was 31.2 percent ($11,578) of that of manufacturing; for other services, it was 62.2 percent ($23,087) of the manufacturing wage (ibid.).

Luzerne County resembled Lackawanna County more than it did other former anthracite mining counties. The average manufacturing pay in Luzerne County was $40,511, the highest of any of the former anthracite mining counties. Between 1997 and 2007, manufacturing employment in Luzerne County fell by 26.3 percent, the same as in Lackawanna County. This was a drop of 6,367 jobs, for a loss of total wages in the sector of $257,933,537 compared to what the wage total would have been had manufacturing remained stable in employment between 1997 and 2007. Had it not been for the drop in employment, the manufacturing wage would have totaled $2,586,609,100, rather than $727,373,000. In 1997, the ratio of manufacturing jobs to retail jobs stood at 1.2 to 1, by 2007, this had fallen to 0.95 to 1, indicating that manufacturing had been displaced as the largest employment sector in the county (ibid.). This was a result of a rapidly declining manufacturing sector coupled with a rapidly expanding retail sector and a rapidly expanding health care and social assistance sector. In 1997, the ratio of manufacturing to health care and social assistance stood at 3 to 1. By 2007, it had fallen to 0.77 to 1. The average pay of health care and social assistance was 85.4 percent ($34,585) of manufacturing (ibid.).

Northumberland County's average manufacturing pay was $37,417 in 2007. The number of workers in manufacturing fell by 22.6 percent from 1997 to 2007. Manufacturing, however, continued to be the largest category of employment throughout the decade, unlike in Luzerne and Lackawanna Counties. Nonetheless, other sectors were growing while manufacturing was contracting. Manufacturing lost 1,764 jobs between 1997 and 2007. This resulted in a loss of possible income in manufacturing of $66,003,588, which would have taken income from manufacturing in Northumberland County to $292,303,588.

The manufacturing to retail sector ratio in Northumberland County declined from 2.1 to 1 to 1.8 to 1, a slight decrease. The retail wage on average was 54.2 percent ($20,294) of that in manufacturing. The change in the ratio of manufacturing to health care and social assistance was somewhat greater, shifting from 4.7 to 1 to 1.6 to 1 between 1997 and 2007. The average wage in the health care and social assistance section was $26,927, or 72 percent of the average in manufacturing (ibid.).

The manufacturing to real estate ratio indicated a degree of stability. In 1997, it was 33.8 to 1, and in 2007 it was 31.3 to 1, indicating a slight growth in positions in real estate compared to manufacturing in Northumberland County. The ratio of manufacturing employment to accommodation and food also was fairly stable. In 1997, it stood at 4.8 to 1, and in 2007 it was 4.1 to 1. The wage was especially low in the accommodation and food sector. The average payroll amount was $9,676, which was just 25.9 percent of the manufacturing average payroll amount. Manufacturing compared to other services shifted from 14.1 to 1 to 7.5 to 1, indicating a strong growth in this category relative to the decline in manufacturing. Other services paid $15,378 on average, which was 41.1 percent of the average manufacturing wage (ibid).

Finally, in Schuylkill County the average manufacturing wage was $38,960. Between 1997 and 2007, employment in manufacturing declined by 28.9 percent, the second largest decline in the former anthracite fields. This amounted to 4,159 lost manufacturing jobs. The possible income loss in 2007 was $162,034,640, which was second only to the possible losses in Luzerne County. Had these jobs not been lost, the total wage in manufacturing in 2007 would have been $559,853,640, rather than the $397,819,000 that was obtained. Retail grew relatively slowly compared to the decline in manufacturing. In 1997, there were 2.1 manufacturing jobs for every retail position. In 2007, the ratio was 1.7 to 1. Retail paid 48.9 percent ($19,066) of what manufacturing did in 2007 (ibid.).

Health care grew relative to the level of manufacturing. In 1997, the ratio of manufacturing jobs to health care jobs was 4.7 to 1, by 2007, this figure, was 1.3 to 1. The two categories had drawn just about even in terms of the distribution of jobs in Schuylkill County. Health care paid 78.7 percent ($30,661) of what manufacturing did in 2007 (ibid.).

There was a substantial shift in the manufacturing to real estate jobs ratio. It declined from 56.4 to 1 to 29.3 to 1. At an average of $21,862, real estate wage was 56.1 percent of the manufacturing wage. There was also a change in the manufacturing to food and accommodation sector. The ratio stood at 4.9 to 1 in 1997 and 3.4 to 1 in 2007 (ibid.). Accommodation and food had the lowest average wage in Schuylkill County, as it did in all of the other former anthracite counties. In 2007, that average was $11,500, which was 29.5 percent of the manufacturing wage (ibid.).

The manufacturing to other services ratio fell sharply from 1997 to 2007. In 1997, it was 18 to 1, and in 2007 it was 8.6 to 1. Other services paid only 49.5 percent ($19,299) of the manufacturing wage in Schuylkill County.

The following data show how other sectors, generally lower-paying, have increased across the time span from 1997 to 2007. In all of the former anthracite counties, the fastest growing sector was health care. In Carbon County, there were 4.17 health care jobs in 2007 to every 1 such job in 1997.

In Lackawanna County, the ratio was a somewhat lower 3.57 to 1, but this far exceeded the growth in other sectors. In Luzerne County, the health care sector had 2.94 jobs in 2007 for every 1 that existed in 1997. Schuylkill County's health care job growth from 1997 to 2007 was somewhat lower than in the other with a ratio of 2.63 to 1. The slowest growth in health care employment took place in Northumberland County, where the ratio was 2.33 to 1 (ibid.).

The pattern of change in employment was uneven for the retail sector across the five former anthracite counties. In Luzerne County, it was stationary, with as many retail jobs in 2007 as there had been in 1997. The retail sector increased slightly in Carbon and Lackawanna Counties, with ratios of 2007 to 1997 employment at 1.1 to 1 and 1.08 to 1, respectively. Given these small increases, the retail sector only grew compared to manufacturing in a relative sense, as manufacturing contracted substantially in these counties. In Northumberland and Schuylkill Counties, the retail sectors actually contracted. The ratio for Northumberland County was 0.89 retail jobs in 2007 for every 1 job in 1997, and in Schuylkill County, it was 0.85 retail jobs in 2007 for every 1 job in 1997 (ibid.)

In four of the five counties, the real estate sector grew from 1997 to 2007. The exception was Northumberland County, where it contracted. There were 0.83 real estate jobs in 2007 for every 1 real estate job in 1997 in Northumberland. The highest growth in real estate jobs took place in Carbon County, where there were 1.45 jobs in 2007 for every 1 in 1997. Schuylkill, Luzerne, and Lackawanna Counties were next with respective ratios of 1.37 to 1, 1.35 to 1, and 1.22 to 1 (ibid.).

Accommodations and food went virtually unchanged from 1997 to 2007 in Luzerne County (1.08 to 1) and Schuylkill County (1.01 to 1). It declined in Northumberland County, where the ratio was 0.88 to 1, and had its greatest increases in Carbon County, with a ratio of 1.25 to 1, and Lackawanna County, where the ratio was 1.16 to 1 (ibid.). The increase in Carbon County was not unexpected given its growing tourism industry.

Other services expanded in four out of the five counties. Lackawanna County was the sole exception, where the 2007 to 1997 ratio was 1.08 to 1, indicating stagnation. The largest increase was in Carbon County, with a ratio of 1.92 to 1, followed by Schuylkill and Northumberland Counties, with respective ratios of 1.49 and 1.47 to 1. Luzerne's was 1.32 to 1 (ibid.).

The 2000 U.S. Census and the American Fact Finder (2010) supply different sectoral data for comparisons in 2000 and 2008 than does the U.S. Bureau of the Census Selected Characteristic by Economic Sector. Looking at the data for each county and the patterns of change from 2000 to 2008 from the U.S. Census and the American Fact Finder, plus the 1990 data on manufacturing, we observe the following. Persons working in the construction sector made up 8.5 percent of the workforce in 2000 and 9.6 percent in 2008, indicating a slight increase in importance in this sector over time. As noted previously, the manufacturing sector in Carbon County contracted sharply from 2000 to 2008, falling from 22.1 percent of the workforce to 15.0 percent. From 1990 to 2000, manufacturing employment declined from 32.4 to 22.1 percent in 2000, a drop of 10.3 percent, steeper than from 2000 to 2008 (U.S. Bureau of the Census 1990). There was a slight drop in employment in wholesale trade from 3.1 to 2.3 percent. On the other hand, retail trade increased as a share of Carbon County's workforce from 12.1 to 13.8 percent between 2000 and 2008. In 2000, transportation and warehousing, and utilities workers made up 6.7 percent of the workforce. In 2008, they were 7.4 percent. Information workers remained unchanged at 2.5 percent of all employees in 2000 and 2008. There was a slight decrease in the percentage of employees working in the finance, insurance, real estate, rental, and leasing sector. Between 2000 and 2008, the percentage fell from 4.2 to 3.9. A very slight increase took place for employment in the professional, scientific, management, administrative, and waste management services. In 2000, 5.8 percent of all employees were in this sector; in 2008, there were 6.1 percent.

Educational, health, and social services jumped substantially as a percentage of Carbon County's workforce, increasing from 19.5 percent (the second largest category behind manufacturing in 2000) to 22.2 percent (the largest employment sector in the county in 2008). Arts, entertainment, recreation, accommodation, and food service changed between 2000 and 2008. In the former year, it stood at 6.5 percent and in the latter, it was 7.6 percent. Other services (except public administration) shifted by a small amount from 4.1 percent in 2000 to 4.7 percent in 2008. Finally, public administration was virtually unchanged from 2000 to 2008. In 2000, it was 3.8 percent and in 2008 it was 3.7 percent. The data for Carbon County, with the exception of workers in the manufacturing sector and in the education, health, and social services sector showed a great deal of stability between 2000 and 2008 (U.S. Bureau of the Census 2000; American Fact Finder 2010).

Workers in the construction sector increased slightly from 5.2 percent in 2000 to 5.6 percent in 2008 in Lackawanna County. The manufacturing sector shrank between 1990 and 2000, declining from 21.8 percent in 1990 to 15.7 percent in 2000, a shift of 6.1 percent. It further dropped from 15.7 percent in 2000 to 12.2 percent in 2008. In neither 2000 nor 2008 was manufacturing the largest employment sector in Lackawanna County. It, however, was in 1990 (U.S. Bureau of the Census 1990). Wholesale trade was virtually unchanged over the period under consideration. It stood at 3.6 percent in 2000 and was estimated to be 3.5 percent in 2008. The retail trade sector employed 13.4 percent of the workforce in 2000 and 13.5 percent in 2008. Also virtually unchanged between 2000 and 2008 was employment in the transportation, warehousing, and utilities sector, where employment was at 5.5 percent in 2000 and 5.7 percent in 2008. The information sector employed 2.8 percent of workers in 2000 and 2 percent in 2008. Employment in finance, insurance, real estate, and rental and leasing was virtually constant between 2000 and 2008, with employment in the sector 6.6 percent in the former and 6.7 percent in the latter year. Employment in the professional, scientific, management, administrative, and waste management services grew from 6 percent in 2000 to 6.7 percent in 2008. In 2000, the largest employment sector in Lackawanna County was educational, health, and social services, with 24.2 percent of all county workers. It was also the largest sector in 2008, with 25.9 percent. In Lackawanna County, the growth in this sector was much smaller than in Carbon County for a comparable time period.

There was an increase of 0.3 percent in arts, entertainment, recreation, accommodation, and food services in Lackawanna County. In 2000, employment in this sector was 6.7 percent of the county's workforce and in 2008 it was 7 percent. Other services (except public administration) were 5.4 percent of the workforce in 2000 and 4.3 percent in 2008, while public administration was 4.5 and 5.8 percent respectively in 2000 and 2008. Lackawanna County's employment patterns by sector were more stable, overall, than Carbon County's. While the manufacturing sector decreased in both counties, the drop was much greater in Carbon County than in Lackawanna County in 2008 (U.S. Bureau of the Census 2000; American Fact Finder 2008, 2010).

The construction sector was unchanged at 5.9 percent of the workforce in both 2000 and 2008 in Luzerne County. Manufacturing contained 23.4 percent of Luzerne County's workforce in 1990 (U.S. Bureau of the Census 1990). Between 1990 and 2000, this percentage fell by 6.8 percent to 16.6 percent. Manufacturing sector employment decreased by another 3.1 percent, from 16.6 to 13.5 percent between 2000 and 2008. This was a little less than the drop of 3.5 percent observed for Lackawanna County, but, again, was smaller than the decline in Carbon County. Employment in wholesale trade was stable at 4.2 percent in both years. The retail trade sector grew slightly from 13 to 13.8 percent, which was approximately the same as Carbon and Lackawanna Counties' employment in this sector in 2008. The transportation, warehousing, and utilities category was 5.8 percent in 2000 and 5.3 percent in 2008. Information employment had a slight decrease, dropping from 5.8 to 5.3 percent between 2000 and 2008. In 2000, 5.8 percent of the workforce was in the finance, insurance, real estate, and rental sector. In 2008, this sector made up 6 percent of all employees in Luzerne County, a figure comparable to Lackawanna but greater than Carbon County's percent. In Luzerne County, 6.2 percent of workers were employed in the professional, scientific, management, administrative, and waste management sector in 2000. In 2008, this was 7.3, an increase of 1.1. By far, the largest employment sector in 2000 and 2008, as in Lackawanna County, was the educational, health, and social services sector, with 21.5 percent of all workers in 2000 and 23.2 percent in 2008. Arts, entertainment, recreation, accommodation, and food services employed 7 percent of county workers in 2000 and 7.9 percent in 2008. The percentages for other services (except public administration) were 4.4 and 4.1, and for public administration they were 5.4 and 5.2 percent in 2000 and 2008 respectively (U.S. Bureau of the Census 2000; American Fact Finder 2010).

In Northumberland County, 5.8 percent of the workforce in 2000 was in the construction sector; by 2008, that number had increased slightly to 6.6 percent. As in all of the other counties, manufacturing slid, between 1990 and 2008, decreasing from 27.7 percent in 1990 to 21.8 percent in 2000 and 18.1 percent in 2008 (U.S. Bureau of the Census 1990, 2000; American Fact Finder 2010). This continuing decline moved manufacturing into a clear second place among employment sectors in 2008. Retail trade employment declined slightly from 13.3 percent of the labor force in 2000 to 12.5 percent in 2008. Transportation, warehousing, and utilities increased from 6.0 to 6.6 percent of the workforce between 2000 and 2008. Information slightly more than doubled as a percentage of Northumberland County's labor force from 1.3 to 2.7 percent. Finance, insurance, and real estate fell from 3.6 to 2.7 percent. Professional, scientific, management, administrative, and waste management services made up 3.7 percent of Northumberland's workforce in 2000. In 2008, 3.9 percent were in this category. Educational services, health care, and social assistance increased by 4.1 percent, the largest increase of any sector, between 2000 and 2008. There was a slight increase in employment from 6.4 to 7.1 percent in the arts, entertainment, recreation, accommodation, and food services sector across the time period. Other services fell slightly from 4.4 to 4.1 percent, and public administration also dropped, declining from 6.7 in 2000 to 6.2 percent in 2008 (U.S. Bureau of the Census 2000; American Fact Finder 2010).

In Schuylkill County, construction workers made up 6.6 percent of the labor force in 2000 and 5.0 percent in 2008: a decline of 1.6 percent. The drop in manufacturing was larger. In 1990, the percentage of workers in manufacturing was 33.3 percent, the highest percentage in the five-county former anthracite region (U.S. Bureau of the Census 1990). By 2000, the percentage had fallen to 26 percent, a decline of 7.7 percent, and by 2008, it had dropped to 20.3 percent, a further decline of 5.7 percent, making manufacturing the second largest employment sector in the county. Wholesale trade dropped from 3.6 to 3.2 percent, retail employment increased by 1 percent from 12.3 to 13.3 percent. Transportation, warehousing, and utilities increased from 4.9 to 6.2 percent. Information stayed almost constant. It was 1.5 percent in 2000 and 1.4 percent in 2008. Finance, insurance, real estate, and rental and leasing declined from 3.8 percent of the labor force to 3.3 percent. Professional, scientific, management, administrative, and waste management services stood at 5.2 percent in 2000 and 5 percent in 2008.

Education, health, and social services was the fastest growing employment sector in the county's economy. It increased from 19.5 percent of all workers in 2000 to 22.6 percent in 2008. There was only a slight change in arts, entertainment, recreation, accommodation, and food services. It went from 5.2 to 5.5 percent. Other services were virtually unchanged, standing at 4.7 percent of all employees in 2000 and 4.6 percent in 2008. Finally, public administration increased from 4.7 percent in 2000 to 5.6 percent in 2008.

We next examine the ratios of 2007 to 1997 incomes by sectors identified in the U.S. Bureau of the Census's Selected Economic Characteristics within each county, unadjusted for inflation. In Carbon County, wage growth was highest in manufacturing and real estate. The ratio of 2007 wages to 1997 wages was 1.9 to 1 in each sector. In retail, the 2007 to 1997 wage ratio was 1.4 to 1, in health and social assistance it was 1.3 to 1, in accommodation and food it was 1.5 to 1, and in other services the ratio was a meager 1.1 to 1, indicating a high level of wage stagnation in this last sector (U.S. Bureau of the Census 1997, 2007).

In Lackawanna County, wage growth in the manufacturing sector was 1.3 to 1, lower than that for Carbon County. The change in the retail wage was 1.4 to 1, the same as was found in Carbon County. The health sector ratio was 1.2 to 1 and real estate was at 1.4 to 1, the same as retail. In accommodation, there was a decline in 2007 income compared to 1997 income, with a ratio of 0.7 to 1. In other services, there was no wage growth across the decade (ibid.).

Wage growth was highly consistent across sectors in Luzerne County. Manufacturing, retail, real estate, and accommodation and food all had ratios of 1.4 to 1, and health care and social assistance was 1.3 to 1. Wages in other services contracted from 1997 to 2007. The ratio was 0.8 to 1 in 2007 (ibid.).

Northumberland County's manufacturing wage ratio for 2007 compared to 1997 was 1.4 to 1, as was the wage ratio for accommodation. Retail stood at 1.3 to 1, and two sectors, health and social assistance and real estate, were 1.6 to 1. The wage in other services was stagnant, with a ratio of 1 to 1 (ibid.).

In Schuylkill County, wage growth in manufacturing expressed itself in a 2007 to 1997 ratio of 1.4 to 1, the same as was obtained for Luzerne and Northumberland Counties and not much larger than the 1.3 to 1 ratio found for Lackawanna County. Retail and health care and social assistance stood at 1.3 to 1. Real estate was stagnant with 1 to 1 ratio. The largest wage growth appeared in the accommodation and food sector at 1.5 to 1. Other services contracted, as it had in Luzerne County, with a ratio of 0.8 to 1 (ibid.).

Finally, while in 1997 the retail trade wage in Carbon County stood at 70.5 percent of the county's manufacturing wage, by 2007 it made up only

53.9 percent of the latter. Thus, wage gains in the retail sector had not kept pace with those in the manufacturing sector. Similarly, while in Carbon County in 1997, health care jobs paid 111.6 percent of manufacturing jobs, by 2007, differential patterns of change in manufacturing and health care wages had put health care jobs at only 82.2 percent of the manufacturing wage. The real estate wage was 78.6 percent of the manufacturing wage in Carbon County in 1997, but by 2007, it had increased to 81.9 percent of the average manufacturing wage. The accommodation and food wage dropped dramatically as a percentage of the manufacturing wage in Carbon County over the decade. It fell from 45.1 to 37.6 percent of the manufacturing wage. The other services' average wage also fell dramatically in Carbon County from 78.8 to 48.5 percent of the manufacturing wage (ibid.).

In Lackawanna County in 1997, the average retail wage was 50 percent of the average manufacturing wage. In 2007, this figure stood at 53.5 percent, indicating a slight increase relative to what manufacturing workers were earning. In 1997, health care workers made more than manufacturing workers, earning, on average, 104.3 percent of the manufacturing wage. By 2007, this had fallen to 93.7 percent. The real estate wage in Lackawanna County was 59.5 percent of the manufacturing wage in 1997. By 2007, it was 65.7 percent of the manufacturing wage. The accommodation and food wage was 45.1 percent of the manufacturing wage in 1997. In 2007, it had declined in Lackawanna County to 31.2 percent of the average manufacturing wage. The other services wage was 82.2 percent of the manufacturing wage in 1997. It increased less than did the manufacturing wage between 1997 and 2007. As a result, by 2007 this sector's wage was only 62.2 percent of the average wage for manufacturing (ibid.).

The retail wage in 1997 in Luzerne County was 50.6 percent of the manufacturing wage. In 2007, the retail wage was a slightly higher proportion of the manufacturing wage (51.2%) than it was in 1997. The average health care wage in 1997 stood at 93.3 percent of Luzerne County's average manufacturing wage. In 2007, it had fallen to 85.4 percent of the average manufacturing wage. The real estate wage also fell as a proportion of the manufacturing wage between 1997 and 2007. In 1997, it was 74.4 percent of the average manufacturing wage, but by 2007, it had decreased to 65.7 percent. The accommodation and food sector, which was 28.4 percent of the manufacturing wage in 1997, dropped slightly to 27.9 percent in 2007. As a percentage of the manufacturing wage, the other services sector dropped dramatically in Luzerne County from 90.3 percent of the manufacturing wage in 1997 to 48.3 percent in 2007 (ibid.).

In Northumberland County, the retail sector wage, relative to the manufacturing sector wage, was fairly stable between 1997 and 2007. In 1997, it was 57.4 percent of the manufacturing wage and in 2007 it was 54.2 percent.

The accommodation and food sector wage also remained fairly constant as a proportion of the manufacturing wage, changing from 26.7 to 25.9 percent. The real estate wage increased relative to the manufacturing wage between 1997 and 2007, shifting from 54.7 to 65.2 percent. The health care average wage also increased as a proportion of the manufacturing wage in Northumberland County from 60.8 to 72 percent between 1997 and 2007. On the other hand, other services declined from 54.9 to 41.1 percent (ibid.).

In Schuylkill County in 1997, the retail wage, on average, stood at 54.4 percent of the manufacturing wage. It fell by 2007 to 48.9 percent. The average health care wage also declined as a percentage of the manufacturing wage from 1997 to 2007, dropping from 88.2 to 78.7 percent. The real estate wage on average likewise dropped substantially as a percentage of the manufacturing wage from 1997 to 2007. In 1997, the average wage for real estate was 77.4 percent of the average manufacturing wage, while in 2007 it was 56.1 percent. Other services average wage fell from 85.1 percent of the average manufacturing wage to 49.5 percent, while accommodation and food was relatively stable, increasing from 28.4 to 29.5 percent (ibid.).

Wage growth across sectors, expressed in constant 2007 dollars, was uneven within counties. In Carbon County, the manufacturing wage in 1997 was $27,006, which was $11,827 less than the manufacturing wage in 2007. This difference in wages represented a 43.8 percent increase in the manufacturing sector wage over the decade (ibid.). Recall Carbon County also had the largest contraction in the percentage of jobs in manufacturing from 1997 to 2007. The retail wage in Carbon County was $1,921 more in 2007 than it was in 1997, an increase of 10.1 percent, quite a bit below the percentage increase in the manufacturing sector. In health care and social assistance, from 1997 to 2007, the wage increase was $1,028 or 3.4 percent, for an average of 0.34 percent a year. In the accommodation and food sector, the wage increase was $2,349. This was an increase of 19.1 percent in constant dollars. The other services wage fell by $2,468 or 11.6 percent (ibid.). These data show that in Carbon County, wage growth was not substantial in any sector except manufacturing from 1997 to 2007, the very sector that was contracting in employment. The expanding employment sectors' wage grew at much slower paces or actually fell in real terms compared to manufacturing over the decade.

In Lackawanna County, the 1997 manufacturing wage stood at $37,262 expressed in 2007 dollars. This was $160 or 0.43 percent less than the 2007 wage. In effect, the manufacturing wage was basically unchanged when examined in constant dollars, a far cry from the situation in Carbon County. Thus, not only were manufacturing jobs being lost in Lackawanna County but the manufacturing wage was stagnant between 1997 and 2007. The retail trade wage grew in constant dollars by $1,241 between 1997 and 2007.

This was an average increase of 6.7 percent over the decade. In 2007 dollars, the health care wage was $38,858 in 1997. This was $4,099 less than the 2007 wage in this sector. The real wage in the real estate sector increased by $2,205 over the ten years, a 9.9 percent increase. The accommodation and food real wage dropped by $8,809 or 43.2 percent, the largest decrease, by far, of any sector in Lackawanna County. The other services wage fell by $7,946, a decline of 25.6 percent (ibid.). All of these data for Lackawanna County show an economy in deep distress between 1997 and 2007. It seems to have been helped very little by Governor Ridge's economic development strategy.

Luzerne County's manufacturing wage, expressed in 2007 dollars, in 1997 was $37,161, which was $3,350 less than the 2007 wage in this sector. This number represented an increase of 9 percent over the decade, more than was found for Lackawanna County but less than that observed for Carbon County's manufacturing sector. While the manufacturing wage increased from 1997 to 2007, so too did the retail sector wage. On average, retail workers made $1,964 more in 2007 than they did in constant dollars in 1997. This was an increase of 10.4 percent, slightly more than the percentage increase in the manufacturing wage. Health care, the fastest growing sector in Luzerne County, experienced a decline of $4,273, which equated to an 11 percent decline in the real average wage.

The real estate sector wage grew from 1997 to 2007 in constant dollars in Luzerne County. In 2007, the average wage in real estate was $1,747 higher than in 1997, an increase of 6.3 percent. The wage also increased, albeit by a small amount, in the accommodation and food sector. In 2007, the average real wage was $775 greater than it had been in constant dollars in 1997. The average wage in other services, however, decreased by $13,982 in real terms. This was a 41.7 percent decline (ibid.).

The real manufacturing wage increased by $1,482 between 1997 and 2007 in Northumberland County. This represented an increase of 4.1 percent, smaller than Carbon and Luzerne Counties but more than Lackawanna. The retail sector wage, on the other hand, expressed in constant dollars decreased between 1997 and 2007. The decline was $315 or 1.5 percent. The health care and social assistance sector wage also increased in Northumberland County. Workers in this sector averaged $5,095 more than they earned in 1997, a 23.3 percent increase. But despite the increase in Northumberland County's wages in this sector and the decline of Luzerne County's wages, Northumberland County health care workers still lagged behind the average earned in Luzerne County in this sector in 2007 (ibid.).

Real estate workers experienced wage increases in real terms between 1997 and 2007 in Northumberland. On average, they made $4,726 or 24 percent more over the ten years, an increase in real terms approximately the size of that for health care and social assistance workers. The average wage of accommodation and food workers increased slightly in real terms, $457, which was approximately a 5 percent increase in real income over the decade. Other service workers lost pay in real terms over the decade. The average real wage fell in this sector by $4,094, for a decrease of 20.7 percent (ibid.).

Schuylkill County manufacturing workers received a real wage increase of $3,570. This is an average increase of 10.1 percent across the decade. Retail workers on the other hand experienced a slight decrease in wages over this period. The average real wage dropped by $199, a fall of 1 percent. Health care and social assistance sector workers on average lost $536 in real wages between 1997 and 2007, a drop of 1.7 percent. Real estate workers, on average, gained $646 in income. This was an increase of 3 percent. Accommodation and food workers received an average increase of $1,449 or 14.4 percent between 1997 and 2007. Other services workers experienced a large drop in wages between 1997 and 2007. Their real income declined by $10,827, a collapse of 35.9 percent (ibid.).

Looking at household, family, and per capita income between 1999 and 2008, expressed in 2008 dollars, using the Consumer Price Index Inflation Calculator (U.S. Bureau of Labor Statistics 2008, 2010), we see that these all fell in real dollars in all of the counties except for Carbon, where median household income and per capita income increased between 1999 and 2008 (U.S. Bureau of the Census 2000; American Fact Finder 2010). In 2008, median household income was $46,727 in Carbon County. In 1999, expressed in 2008 dollars, it was $45,378. This was an increase in real dollars of $1,349, or a ratio of 1.03 to 1. Median family income, however, declined in real terms. It dropped from $54,431 in 1999 to $53,269, a loss of $1,162, yielding a ratio of 0.98 to 1. Per capita income increased from $22,052 in 1999 to $22,806, a gain of $754. The ratio of 1999 to 2008 per capita income was 1.03 to 1. Families were thus slightly worse off in purchasing power in 2008 than they had been in 1999, while households gained a small amount in purchasing power in Carbon County between 1999 and 2008.

In Lackawanna County between 1999 and 2008, real purchasing power for households fell by $2,379. It went from $44,505 to $42,126. The ratio of median household income in 2008 to 1999 was 0.95 to 1. For median family income in Lackawanna County, between 1999 and 2008 there was a drop in real income from $58,089 to $56,808, for a decline of $1,281. The ratio of 2008 median family income to 1999 median family income stood at 0.98 to 1. Per capita income declined from $24,180 to $23,568. The 2008 to 1999 ratio for per capita income stood at 0.97 to 1 (U.S. Bureau of the Census 2000).

Luzerne County exhibited a pattern of income change similar to Lackawanna County, with median household income, median family income, and per capita income all lower in 2008 than they were in 1999, expressed in constant 2008 dollars. Median household income dropped from $43,643 to $42,388, which was $1,255 lower in 2008 than it was in 1999, a ratio of 0.97 to 1. For median family income, the 1999 figure was $56,003, and in 2008 it was $54,216, a decline of $1,787, a ratio of 0.97 to 1. Per capita income also fell. In 1999, it was $23,557, while in 2008, it was $23,028, a difference of $529, for a ratio of 0.98 to 1 (ibid.).

Median household income, expressed in 2008 constant dollars, dropped in Northumberland County from $40,597 to $39,266 between 1999 and 2008, a loss of $1,331. The ratio of 1999 median household income to 2008 was 0.97 to 1. Median family income declined from $51,113 to $49,696, a difference of $1,417, producing a ratio of 0.97 to 1. Per capita income also declined from 1999 to 2008. In 1999, it was $21,309, and in 2008, it was $20,502, a difference of $807. The ratio was 0.96 to 1 (ibid.).

All three measures of income fell in Schuylkill County between 1999 and 2008. Schuylkill County, however, had the smallest drop of any county seeing declines in the former anthracite region. Median household income in 2007dollars in 1999 was $42,258, and in 2008, it was $42,041. This is a difference of $217, yielding a ratio of 0.99 to 1—a small change. Median family income dropped from $53,346 to $53,111, a change of $235. The ratio of 1999 dollars to 2008 dollars again was 0.99 to 1, again indicating a small change. Per capita income fell from 1999 to 2008 by $902, from $22,267 to $21,365. The ratio was 0.96 to 1, a slightly larger change than we saw in the previous measures of income loss in Schuylkill County (ibid.).

Given all of these data, it is clear that there was a substantial trade down in employment between 1997 and 2007. Higher-paying manufacturing jobs gave way to jobs in lower-paying economic sectors, both in an absolute and a relative sense. It is worth noting that Luzerne County's manufacturing wage of over $40,000, the highest in the region, was below Pennsylvania's average manufacturing wage. Matters became even more complicated as unemployment increased with the Great Recession. Time after time, attempts to revitalize the former anthracite coal regions have been marginally successful at best, and failures at worst. The region has undergone successive waves of deindustrialization and was particularly hard hit by the Great Recession. As a result, it continues to lag behind the rest of Pennsylvania on most measures of economic and social well-being, and continues to lose population as a result of the dying off of an aging population and the out-migration of younger workers. The median age in Pennsylvania in 2000, according to the U.S. Census, was thirty-eight, the highest of any state in the middle-Atlantic region. The five counties in the former coalfields were higher than this, with Carbon County at 41.3, Lackawanna 40.3, Luzerne 40.8, Northumberland 40.9, and Schuylkill 40.9. The percent of persons sixty-five and older in each of these counties, respectively, was 19.8, 19.5, 19.7, 19.0, and 19.9 percent, compared to an average in the commonwealth of 15.6 percent.

Businessmen and women see several advantages to the shift away from manufacturing that has been taking place on a long-term basis since the 1980s. First, it provides for a diversification of the economy, which could offer a modicum of protection in the event of a historically "normal" downturn in the economy. This change, however, provided little protection from the ravages of the Great Recession, the effects of which the region is currently still experiencing. Second, the loss of higher-paying manufacturing jobs puts downward pressure on the average wage throughout other economic sectors in the former anthracite coalfields, thus cheapening the average cost of labor for investors. Third, entrepreneurs hope that the decline in manufacturing will weaken the power of the unions throughout the former anthracite coalfields, making it easier to effect changes in the political system to benefit corporations and individual investors.

The downside of the loss of manufacturing jobs and the higher wages they pay is a drop in purchasing power throughout the region, a classic dilemma for capitalists, each of whom wants the wages of his or her employees to drop to a minimum, but also wants the wages of other employers' workers to increase. While wages have experienced real growth from 1997 to 2007 in most economic sectors in the former anthracite counties, the rate of increase, especially adjusted for inflation, has not been large year to year, and the changing structures of employment have actually led to decreases in family and per capita income, as well as in household income, except in Carbon County, for the last two years.

With the hope that trade union power is eclipsed in the former coalfields, capital believes that it can push back against the state to weaken the advances labor has made in terms of support for public programs. It is to be expected that, in the face of a massive economic downturn, not only will capital seek concessions on wages and benefits from their own employees, but capital also will begin to demand concessions from the state on selected taxes and public expenditures.

Public employee unions are relatively strong in the former anthracite region. In 2008, 78 percent of public sector workers in Scranton-Wilkes-Barre were organized, compared to 9.1 percent of private sector workers in the same MSA (Hirsch and Mcpherson 2010). This is the only region in the coalfields for which data on unionization rates are available.

In an effort to divide organized workers, capital has been portraying public employees' wages and benefits as being far out of line with what the private sector can furnish. It remains to be seen how successful this campaign against public sector workers and their unions will be in the coal counties, and whether capital's propaganda and political assault on public sector workers will split the solidarity within the area's working classes.

With an already declining economy, the problems of which were exacerbated by the Great Recession, a decreasing and aging population, and a shrinking private sector labor movement (although the public sector labor movement has grown significantly), the institutions that once gave the anthracite communities high levels of social solidarity and social vitality are beginning to whither.

This has led to political fragmentation and isolation, and a reduced capacity to cope with continuing decline. The working population stands at the mercy of the existing centers of capital operating in the region, even though they, the workers, still control substantial portions of the local state apparatus. Such political control has not translated into solid economic benefits. The anthracite coal regions are low-wage economies with few opportunities for upward mobility. The major opportunity for upward economic mobility comes with migration out of the region, which has been occurring on a continuing basis since the coal industry began its long-term decline.

Most of the former anthracite counties lost population between 2000 and 2009, as they had over the past several decades, with an aging population dying off and younger people migrating elsewhere in search of better-paying jobs than are currently available or likely to be in the future. While population increased in Pennsylvania from 2000 to 2009 by 2.6 percent, only one of the five former anthracite counties exhibited any growth. That was Carbon County, which grew by 8.6 percent, perhaps a reflection of its location in the Pocono Mountains, a resort, commuter, and retirement center in the region, and its pattern of wage growth. Lackawanna County's population decreased by an estimated 2.1 percent, Luzerne County's by 2.0 percent, Northumberland County's by 3.4 percent, and Schuylkill County's by 2.3 percent (U.S. Census 2010). This followed on decreases from 1990 to 2000, except for Carbon County (U.S. Census 2000).

The coal regions, as we have documented, are economically distressed to a greater extent than is the Commonwealth of Pennsylvania taken as a whole. These regions also are distressed in terms of physical well-being. They have higher-than-average mortality rates compared to the rest of the commonwealth. Out of sixty-seven counties, Lackawanna ranks fifty-fifth in mortality, Luzerne ranks number sixty, Northumberland is sixty-first, Carbon is sixty-second and Schuylkill is sixty-fourth. The higher the rank, the worse is the rating. These counties do a bit better on measures of morbidity, although they all are in the bottom half of the sixty-seven counties. Schuylkill County ranks fortieth, Carbon ranks forty-first, Lackawanna is forty-fourth, Northumberland is forty-fifth, and Luzerne is fiftieth (University of Wisconsin Population Health Institute 2010, 6–7).

The University of Wisconsin's Population Health Institute also reported on county health factor ratings in Pennsylvania. They used four measures. The first was health behaviors, which included measures of smoking, diet and exercise, alcohol use, and risky sexual behavior. The second, clinical care, included measures of access to care and quality of care. The third were social and economic factors. These included measures of educational attainment, employment, income, family and social support, and public safety. The final measure was the physical environment, which examined environmental quality and the quality of the built environment (ibid., 8).

In general, these indicators show that health factors are poorer in the former anthracite fields than they are elsewhere in Pennsylvania, which helps to explain the morbidity and mortality rankings. On health behaviors, of the five anthracite counties, Lackawanna County ranked highest at twenty-seventh. It was followed by Luzerne County, with a rank in the bottom half of the state, at forty-second. Northumberland was next at fifty-ninth, then Schuylkill at sixty-first, and Carbon at sixty-fourth (ibid., 8–9).

In terms of access to and quality of health care (clinical care), Lackawanna County again was in first place among the five anthracite counties, coming in at thirteenth place in the commonwealth. Luzerne County once again was second, with a rank of sixteenth. Both were in the top one-third of all counties in the commonwealth. Northumberland County was thirty-fourth in Pennsylvania, Schuylkill was forty-ninth, and Carbon fifty-seventh (ibid.).

On social and economic factors, the ordering of counties differs from the previous two scales. Lackawanna County was still ranked the highest at thirtieth place. It was the only anthracite county to rank in the upper-half of all Pennsylvania counties. Luzerne County remained in second place, ranking forty-third, which placed it in the bottom half of all counties. Carbon County no longer had the lowest ranking. Instead, it was third among the former anthracite counties at forty-sixth place in the overall county rankings. Schuylkill was next at fiftieth and Northumberland County last at fifty-eighth, nine spots from the bottom (ibid.). These rankings provide further documentation of the relative socioeconomic advantage the former northern fields have over the former central and southern fields.

In terms of physical environment, Carbon County had the top spot among former anthracite mining counties at fourteenth. Northumberland was eighteenth. These rankings put them in the top one-third of all sixty-seven Pennsylvania counties. Luzerne County ranked thirty-fourth, Lackawanna was thirty-sixth, and Schuylkill ranked fiftieth (ibid.).

In sum, what the anthracite industry has left behind is a befouled landscape in most of the counties where the industry operated; a locale that has been unable to sustain high levels of economic growth; a population that has significant health problems and a profile of higher-than-average health risks;

and a region where one has little hope for a prosperous and healthy future. Successive generations have tried to turn around, on a permanent basis, the former anthracite region. Redevelopment policies such as tax expenditures, subsidized industrial parks, promises of a weakening private sector labor movement, and the like all failed, and the prognosis is not good that a future generation will succeed where others could not. These policies repeatedly have been tried since the collapse of coal, but they have not been successful in anchoring the region's fragile economy, let alone develop a viable manufacturing base.

In all likelihood, the coalfields will remain a source of cheap labor, a region, once again, constituted as a reserve army of labor, ready to be pulled into the economy when times are good and to be shed just as readily when times are bad. The real danger, however, is that the manufacturing jobs that have been declining will disappear for good and never be replaced, leaving a stagnating community with few resources to exploit in order to rejuvenate itself.

Just as we have found variations in consequences for the former anthracite counties as a result of the end of coal and a second and third wave of deindustrialization of manufacturing, variations occur in the effects of these processes on the cities in the coalfields. Northumberland County currently has two cities, Sunbury and Shamokin. Lackawanna County has two, Carbondale and Scranton. Schuylkill County has one, Pottsville. Luzerne County has four: Pittston, Wilkes-Barre, Nanticoke, and Hazleton. Carbon County has none.

The first variation among the cities is in the level of education, specifically, the proportion of those age twenty-five and above who hold a bachelor's degree or higher. There is a clear division between the former northern fields and the former southern/central fields. Sunbury has 10.1 percent of its population with an undergraduate degree or higher, Shamokin 7.6, Pottsville 15.5, Carbondale 14.1, Scranton 19.2, Pittston 15.4, Wilkes-Barre 14.8, Nanticoke 12.4 and Hazleton 12.2. Pennsylvania has 26.7 percent of its population with an undergraduate degree or higher (U.S. Bureau of the Census 2012). None of the former anthracite cities come close to the statewide average.

In all of the cities, non-Hispanic whites are the clear majority of the population. Scranton is 5.5 percent black and 9.9 percent Hispanic. Wilkes-Barre is 10.9 percent black, which is just about the state average of 10.8 percent, and 11.3 percent Hispanic, which exceeds the statewide average of 5.7 percent. Hazleton is 4 percent black and 37.3 percent Hispanic. The sizable Hispanic population in Scranton, Wilkes-Barre, and especially Hazleton often make for tense racial relationships between the Anglo population and Hispanics. Pittston and Nanticoke are not nearly as diverse as those three cities. Pittston is only 1.9 percent black and 2.7 percent

Hispanic, while Nanticoke is 1.6 percent black and 2.6 percent Hispanic. Carbondale resembles these cities in that it is 1 percent black and 3 percent Hispanic. Pottsville is 3.1 percent black and 2.5 percent Hispanic. Sunbury is 2.8 percent black and 6.7 percent Hispanic. Finally, Shamokin is 0.6 percent black and 2.6 percent Hispanic. In sum, blacks and Hispanics tend to be more concentrated in the larger cities in the former anthracite counties that made up the northern fields than in the cities of the southern and central fields (U.S. Bureau of the Census 2010) and the smaller cities of Luzerne and Lackawanna Counties.

The occupational composition of these cities differs within and across regions, reflecting the degree of uneven development within the former anthracite region as a whole (see Smith 2008 for a comprehensive discussion of uneven development as an aspect of capitalist systems of production).

In Pennsylvania, 35.5 percent of the workforce is in "management, business, science, and arts occupations." This is considerably more than what we find in the two Lackawanna County cities, where Scranton has 26.5 percent and Carbondale has 22.3 percent. In Luzerne County, Pittston has 25 percent, Wilkes-Barre 24.7 percent, Nanticoke 22.6 percent, and Hazleton 18.2 percent. All of the Luzerne County cities lag behind the commonwealth average. The two Northumberland County cities, Shamokin and Sunbury, have 17.6 and 20.6 percent respectively. Pottsville, in Schuylkill County, has 28.9 percent in this occupational category. This is the highest percent in the former coalfields, but it still falls fall short of Pennsylvania's average (U.S. Bureau of the Census 2012).

In Pennsylvania, workers in "service occupations" comprise 16.9 percent of all workers. Workers in this sector are more common in Carbondale (26.7%) and Scranton (21.3%). In Pittston, in Luzerne County, 17.8 percent are in this category. Wilkes-Barre has 24.7 percent, Nanticoke has 22.8 percent, and Hazleton has 18.7 percent. All the northern field cities exceed the commonwealth average, showing the northern regional urban economy's dependence on what essentially is a low-wage sector.

Pottsville, in Schuylkill County, has 20.1 percent of its workforce in "service occupations." In Northumberland County, Shamokin has 29.5 percent of its workforce and Sunbury has 25 percent. These data show the degree to which the service sector has come to play a vital role in offering employment to urban residents in the former anthracite coalfields compared to its overall role in Pennsylvania (ibid.).

"Sales and office occupations" have also taken up the slack for declining manufacturing jobs. In Pennsylvania, these jobs make up 25 percent of the entire workforce in the commonwealth. In Carbondale, 23.8 percent of workers fall into this category, and in Scranton 27.9 percent do, more than the Pennsylvania average. In Pittston, the total in these positions is 27.7 percent, while in Wilkes-Barre it is 26.9 percent—the largest employment category in the city. Nanticoke is 29.3 percent and it again is the highest percentage in the city. Hazleton has 25 percent (ibid.). For the southern/central counties' cities, Pottsville in Schuylkill County has 27.7 percent in "sales and office occupations," while Northumberland County's two cities, Shamokin and Sunbury, respectively, have 22.3 and 24.6 percent (ibid.).

"Natural resources, construction, and maintenance occupations" account for 8.7 percent of all Pennsylvania workers. In Carbondale, this figure is 9.2 percent, in Scranton 8.0, Pittston 8.1, Wilkes-Barre 5.6, Nanticoke 7.8, and Hazleton 8.2 percent. Pottsville, in Schuylkill County, is 7.6 percent and Shamokin is 12.4 percent. The total is 5.3 percent in Sunbury (ibid.).

"Production, transportation, and material moving occupations" make up 13.8 percent of Pennsylvania's employment. In Carbondale, it is 18.1 percent, Scranton 16.3, Pittston 21.4, Wilkes-Barre 18.2, Nanticoke 17.4, Hazleton 29.3. Pottsville is 15.6, Shamokin is18.2, and Sunbury is 24.8 percent (ibid.).

Overall, in looking at these data on occupations, there does not appear to be a systematic patterning by region for most employment categories. Variations in cities are path-dependent based on the cities' own individual histories and the choices made in redeveloping local economies after the collapse of anthracite and after subsequent waves of deindustrialization swept over the region.

In Pennsylvania, the top three industry sectors of employment are "educational services, health care, and social assistance" (25.2%), manufacturing (12.8%), and retail trade (11.6%). With respect to industry sector, Carbondale's greatest concentration of workers falls into "educational services, health care, and social assistance"—22.1 percent of all workers are employed in this sector. Manufacturing is second at 16.4 percent, and retail trade is third at 12.3 percent.

Scranton deviates slightly from this pattern. "Educational services, health care and social assistance" is first at 28.7 percent, retail trade is second at 13.2 percent, and manufacturing is third at 10.8 percent.

In Luzerne County, Pittston follows the same pattern as Scranton. Educational services, health care, and social assistance is first with 24 percent, followed by retail trade with 14.1 percent, and manufacturing with 10.6 percent. Wilkes-Barre has 23.6 percent of its resident workers employed in educational services, health care, and social assistance; 14.6 percent in retail trade and 12 percent in manufacturing. In Nanticoke, 23.5 percent are in educational services, health care, and social assistance; 18.8 percent are in retail trade; and 11.6 percent are in manufacturing. Hazleton deviates from the pattern in that it continues to be a manufacturing city. Manufacturing has the largest concentration at 22.4 percent, followed by educational services, health care and social assistance at 18.4 percent, and retail trade at 13.5 percent.

In Schuylkill County, Pottsville has 30.6 percent in educational services, health care, and social assistance. The second greatest concentration is in retail trade at 16.7 percent, followed by manufacturing at 10.5 percent.

Shamokin has 30.7 percent in educational services, health care, and social assistance; 18.1 percent in retail trade; and 13.5 percent in manufacturing. Sunbury has 28.4 percent in educational services, health care, and social assistance; 17.8 percent in retail trade; and 12.7 percent in manufacturing (ibid.).

Overall, most of the cities in the coalfields differ from the Pennsylvania profile, reversing the importance of manufacturing and retail trade, showing the effect that the loss of manufacturing jobs has had on patterns of employment in the former anthracite cities. Additionally, manufacturing employs proportionally more residents in Pennsylvania than in the former anthracite cities, where energy capital and manufacturing capital have been cycled out of the region at a relatively high rate. These cities are in bad shape. They have seen small businesses close down, plants close, populations shrink and tax bases shrivel. Also, in some cases, crime has become a problem over the past decade.

Carbondale's crime rate has remained steady with respect to both violent and property crime in the past decade. Both categories of crime are lower than the U.S. average in the past decade (city-data 2013a). Violent crime increased in Scranton between 2001 and 2005, then began decreasing except for a slight upward tick in 2009, after which it continued its fall. In no year between 2001 and 2011 did violent crime in Scranton exceed the national average. Property crime tells a different story. Beginning in 2005, property crime began increasing in Scranton, and in every year between 2005 and 2011, except for 2005 and 2007, Scranton's property crime rate exceeded the national average (city-data 2013e).

The data show that, compared to other former mining cities, Wilkes-Barre is an especially violent place and one where property is threatened by crime. In eight of the eleven years between 1999 and 2011, (2000 and 2001 are excluded), the violent crime rate exceeded the national average. In the same period, in nine years the property crime rate exceeded the national average (city-data 2013h). In part, the high crime rate reflects the role Wilkes-Barre has as a center for drug trading in northeastern Pennsylvania and the struggles among various drug groups to dominate local markets. The region's drug trade has been penetrated by Philadelphia and New York City traffickers who contend with each other for dominance in Wilkes-Barre and, by extension, northeastern Pennsylvania. Given the activity of traffickers in Wilkes-Barre, it is possible to infer that there is high demand for their product in and around that city and the northeastern Pennsylvania region.

Nanticoke, whose violent crime rate was below the national average between 1999 and 2006, saw an increase to a rate above the national average in 2007 through 2010, and a drop below the national average in 2011. In addition, between 2005 and 2011, the Nanticoke property crime rate was also above the national average (city-data 2013c).

Hazleton, which is stereotyped as a high crime city, turns out to have had a higher than national average rate of violent crime only in 2010 and 2011, and in no year between 1999 and 2011 did its property crime rate exceed the national average (city-data 2013b). This belies the city's public reputation.

Compared to Wilkes-Barre, Pottsville is a relatively safe place. Only in 2011 did its rate of violent crime exceed the national average, and in no year from 1999 to 2011 did the property crime rate exceed the national average (city-data 2013d).

Data are available for Shamokin between 2005 and 2011. In only two of those years, 2005 and 2006, did the city's rate of violent crime exceed the national average, and in no year did property crime rates exceed the national average (city-data 2013f).

Like Wilkes-Barre, Sunbury is a dangerous place as the risk of being a victim of a violent crime is higher than the national average in every year (data are missing for 2001) between 1999 and 2011. Property crime rates for the same period, however, exceeded the national average in only four years—1999, 2000, 2003, and 2011 (city-data 2013g). Crime data, like economic data, for cities do not appear to be affected by the region of the coalfields. There is as much urban variability within regions as there is between regions. Two cities, Sunbury and Wilkes-Barre, stand out as distinctive, with respect to violent crime. Wilkes-Barre has the additional problem of a high rate of property crime for most of the 1999–2011 period.

In the final chapter, we draw a set of general conclusions based on our consideration of a number of factors measured at the county level. We show that northeastern Pennsylvania's former anthracite region is an effect of the former anthracite industry. The region was constructed to meet the needs of that industry, and its physical ecology still reflects those needs. The current spatial order and social conditions are an expression of the demise of mining and of a second and third wave of deindustrialization. Mining's collapse and later instances of de-industrializations were a natural outcome of the national circulation of capital investment, which always seeks profit maximization. Capital flowed out of the area, leaving behind a vast region of underdevelopment that occurred at different rates in the region. The northern coalfields, being more urbanized and more diverse economically, have not been as marginalized as the former southern and central regions, where urban development was weak and a strong economic bourgeoisie, which had some control over local development in the north, never formed. In the central and southern fields, local development was almost exclusively the result of decisions made by capitalists residing outside of the area in Philadelphia, New York, and Baltimore.

Chapter 8

Conclusions

We have drawn on a large number of existing histories to tell a story that contextualizes the current economic plight of the former anthracite coal region of northeastern Pennsylvania. This region, which was central to the U.S. economy for a period of almost seventy years, has been on a long-term decline since the end of World War I. Pennsylvania's anthracite region, to say the least, has had a turbulent economic history. Economic instability has been more the norm than the exception. The coal industry especially was marked by vicious boom-and-bust-cycles, many of which were caused by continuing overproduction and under consumption, given the competitive nature of the industry, until the formation of the Morgan cartel in the late nineteenth century. This cartel was able to impose discipline on producers across the fields. Other crashes were caused by financial panics exogenous to the industry. The presence of a strong and relatively effective cartel created a greater commonality in the working conditions of the miners, which enabled them, in turn, to form a union that could impose more discipline across the coalfields, thereby minimizing the likelihood that labor actions in one section would be undermined by workers continuing to produce in another. There still were splits in the labor force by region, but these were minor after 1902 compared to what had gone before.

The cyclical nature of production wreaked havoc on miners and their families' levels of living. It led to unstable employment as during times of high production, the capitalists drew down on the large reserve army of workers it had assembled and during times of low production, workers were pushed into the ranks of the unemployed or semi-employed. Miners' wages also were unstable, with workers forced to take deep cuts during periods of crisis. In order to survive, mineworkers often deployed their children into the labor market, with boys going to work in and around the mines and young girls going to the mills or into domestic service. Wives sometimes took in laundry or sewing or, if the house was big enough, lodgers. Lacking access to education, the children virtually were destined to repeat their parents' situations.

When mining ended in the late 1950s, it left behind thousands of jobless men. With nothing to do, they idled away their hours hanging around street corners, taverns, pool halls, and the like. Many, giving up hope of ever finding stable employment in the region, moved to the Great Lakes, to Philadelphia's industrial suburbs, or to the New York metropolitan area. A good proportion of the unemployed engaged in long-distance commutes, living where they worked during the week and commuting home on weekends, as many families stayed behind rather than relocate. This pattern of migration resulted in a transformation of gender roles, as women took over management of family budgets, child-rearing, and other decision-making in the family. Even in cases where men did not migrate but remained behind unemployed, the increasing employment of married women outside of the household produced an undermining of traditional patriarchy in the minefields (Dublin and Licht 2008).

Marginal employment opportunities even affected those with college degrees. To find jobs consistent with their educational qualifications at wages that were reasonable, from their perspective, they had to move elsewhere, like the blue-collar workers were doing. A significant amount of human capital was lost to the region in these out-migrations of blue-collar and white-collar labor. Population losses also were associated with the second and third waves of deindustrialization in all of the former mining counties except Carbon County, which benefited from the expansion of communities in the Pocono Mountains and the growth of employment opportunities in the Lehigh Valley.

In order to protect their investments with the loss of coal, the local political elites, in conjunction with local bourgeoisie, launched a number of industrial drives, calling on the local population to contribute money out of their own pockets in order to attract new employers. These efforts were marginally successful for a relatively short period of time. Few of the industries that were attracted at this time still remain as major employers. Where the first drive to re-industrialize seemingly was most successful was in attracting investors in garment plants that were fleeing unionized workforces in Philadelphia and New York. Many of these plants were small operations that subcontracted work from major manufacturers. As was taking place among the remnants of the mining industry, organized crime penetrated the garment industry bringing large-scale corruption with it.

The Scranton/Pittston/Wilkes-Barre/Hazleton area has a long and sordid history of organized crime, dominated first by the Irish and, later, by Sicilians. Organized crime runs gambling operations, prostitution, loan-sharking, protection rackets, and has penetrated legitimate businesses, state functions, and some unions in the area. It exists with the tolerance, if not protection, of local political figures and local police forces, many of which have long existed in symbiotic relations with organized crime. But organized crime has not been the only source of corruption in the region. There were (and continue to be) examples of freelancing public officials who have made use of their offices for private gain. Recently there have been a host of indictments in Luzerne County, including two judges who were charged with sending juvenile "offenders" to a private correctional facility in exchange for cash, a county commissioner and a former state senator charged with corruption in Luzerne County, and two major county officials charged with corruption in Lackawanna County in 2011.

The drive to unionize the garment industry brought workers into direct conflict with the members of organized crime, who had invested in the industry, as well as with regular employers. The International Ladies' Garment Workers' Union (ILGWU) eventually succeeded in gaining recognition in the industry after a bitter strike, and it went on to become a powerful political force in the region. It had an especially close relationship with the area's long-term Democratic congressman Daniel J. Flood. The ILGWU helped to perpetuate the trade union culture that had dominated the region since the mining era. Eventually the garment industry, however, died out as plants were moved into the American South and overseas. There is little left of the industry now except for boarded-up factories and a handful of petty producers. This was the region's second wave of deindustrialization.

Another wave of deindustrialization has been taking place since the end of the twentieth century. The industries brought in to replace garment manufacturing, many of which were nonunion and have managed to stay that way, are now closing down, and production is being relocated elsewhere. These factories paid relatively good wages for the setting they were in, although the wages they paid lagged behind state levels by a considerable amount.

The loss of post-mining manufacturing jobs has put even further downward pressure on local wages, making it even more difficult to eke out a decent level of living, as manufacturing has been replaced by even lower-wage sources of employment. The former anthracite region has traded downward in its employment structures.

One can only wonder about the degree to which the anthracite region's continuing decline is a harbinger of the economic future of other metropolitan regions experiencing the loss of manufacturing jobs due to recession. It very well may be the case that the replacement jobs secured will be of lesser quality than those lost, and may require different skill sets than the workers possess. All across the country, we are seeing massive job losses and a capitalist class that is unwilling to invest in the American manufacturing sector in ways that will preserve and expand the pool of quality employment opportunities for workers. We also are seeing downward pressures on real wages for the working classes. These two trends—a shift downward in the composition of available jobs and low wages—are the primary determinants of lowered standards of living in the former anthracite regions.

The expression of popular, mass anxiety over the economic deterioration of recent years has not resulted in the emergence of a sociopolitical movement on the left. Rather, it has generated a populist movement on the right in the form of the various Tea Parties and support for an anti-labor, anti-welfare state agenda in the Republican Party. This can be seen in the results of the 2010 elections in the former anthracite regions. Lackawanna County was the only one of the five counties that supported Democrats across the board. Lackawanna County supported the Democratic ticket for governor and lieutenant governor, voted for the Democratic Senate candidate, and in those parts of the county that made up the 10th and 11th Congressional Districts, backed the Democratic candidates by substantial majorities.

Luzerne County, despite its overwhelming Democratic registration, voted against the thirteen-term Democratic congressman, Paul Kanjorski, in the 11th Congressional District. The county also voted for the Tea Party-backed Republican candidates for governor and lieutenant governor. Luzerne County voters, however, did support the Democratic candidate for the U.S. Senate, Joe Sestak, over the Republican candidate, Pat Toomey, who was also supported by Tea Party groups and won commonwealth-wide.

Kanjorski's successful opponent was Lou Barletta, Republican mayor of Hazleton. This marked the third time that Barletta had run against Kanjorski. He had achieved a good deal of local notoriety and some national attention for his hardline stance against the presence of undocumented immigrants, principally Hispanics, in his city. He had led the passage of an ordinance called the "Illegal Immigration Relief Act," which, among other things, fined people who rented residences to undocumented immigrants. He ran a campaign organized around his opposition to immigration of undocumented workers, to abortion, and to high government spending. He argued that the United States needed to tighten its belt and live within its means. He also claimed that he was opposed to any cuts in Social Security, to raising the retirement age, to privatization of Social Security, and to personal Social Security savings accounts. He advocated strengthening national security despite the fact that the military budget was one of the major causes of the high debt load being carried by the federal government. Barletta also opposed President Barack Obama's health care reform legislation. To control medical costs, Barletta advocated the development of wellness programs where people would take more personal responsibility for their own health and laws that would permit health insurance to be sold across state lines, thus introducing competition into the insurance market and, ultimately, according to his lights, lowering costs. This was consistent with what Tea Party candidates were arguing in 2010.

Once elected, Barletta aligned himself with the right wing of the Republican caucus, proving himself to be an advocate for dismantling the New Deal and its various emanations. He supported Paul Ryan's proposed budget that would have turned Medicare into a voucher system, thereby destroying Medicare as we know it. He also voted for the "Cut, Cap, and Balance" legislation that was among the most radical proposals to come before the House in some time. It would have reduced Social Security payments over the long-term and cut deeply into other public programs while protecting subsidies to agriculture and oil companies, and kept tax expenditures flowing to corporations and the bourgeoisie. Barletta also received the support of voters in Carbon County, as did Toomey, and the Republican slate for governor and lieutenant governor. Barletta was reelected in 2012.

Northumberland County, perhaps the most distressed of the former anthracite counties, supported Republicans across the board. It backed the Republican candidate, Thomas Marion, in the 10th Congressional District, but not by enough to offset the overwhelming support that his opponent, Christopher Paul Carney, had received in Lackawanna County and the smaller margin he won by in Luzerne County.

Schuylkill County voters backed Republicans for governor and lieutenant governor and for the U.S. Senate. They, however, voted for the Democratic candidate in the 17th Congressional District.

Looking at the results of these elections in 2010, the Tea Party candidates and the conservative Republican, Lou Barletta, did very well in the former anthracite fields, with the exception of Lackawanna County.

In the 2008 Presidential elections, Barack Obama carried Carbon County by a margin of 50 to 48 percent, Lackawanna County by 63 to 37 percent, and Luzerne County by 54 to 45 percent. Northumberland County went to John McCain by a vote of 56 to 42 percent and Schuylkill County by 54 to 45 percent.

The Democratic Party had a harder time in the 2010 gubernatorial elections. Tom Corbett, the Republican who was supported by the Tea Party, swept the former coalfields with the exception of Lackawanna County. Lackawanna County backed the Democrat Dan Onarato.

When President Obama ran for re-election in 2012, he won two of the five former anthracite counties: Lackawanna by 62.7 percent to Mitt Romney's 35.7 percent, which was approximately the same margin of victory he had in 2008, and Luzerne by 51.5 to 46.7 percent. This was a smaller margin of victory than in 2008. President Obama lost Carbon County, which he had won in 2008. Carbon County gave President Obama 46.7 percent and Mitt Romney 52.6 percent. Northumberland and Schuylkill Counties once again supported the Republican candidate by 58.5 percent to 39.2 percent and 55.6 to 42.3 percent respectively.

The capitalists behind the Tea Parties and conservative candidates at all levels realized the importance of maintaining structures of consciousness in the American population and an ideology that preserved and advanced laissez-faire capitalism and the core values that supported it—individual responsibility, a strong work-ethic, antiunion sentiments, small government, and anti-collectivism of any type. These values once were generated on the shop-floors in America's factories and the structures of daily life, but with the decline of industrial production, they no longer could be so closely anchored to the society's production processes. Hence, there has developed a need for supplementary apparatuses, such as right wing policy groups, foundations, and think tanks, to anchor and reproduce a political consciousness aligned with bourgeois needs and interests. Among the means to achieve this is to obscure the linkage between political and economic stances and the class interests of the bourgeoisie so that bourgeois needs and interests appear to be the needs and interests of all humanity for all time.

Major corporations, foundations, think tanks, and the like on the right pumped tens of millions of dollars into state and federal legislative races and into races for governorships with the goal of altering the political consciousness of the American masses to make it seem as if the interests of the bourgeoisie were those of all Americans.

In large part, the bourgeoisie and their allies were successful nationally and in the former anthracite regions, with the exception of Lackawanna County and to a lesser extent Luzerne County. The anthracite counties helped put into office Tea Party-backed candidates for governor and lieutenant governor and for U.S. Senate, as well as the Tea Party-aligned conservative Republican Lou Barletta. All of these politicians have pursued policies at the state and national levels that run counter to the social and economic interests of the populations of the former anthracite regions and which, if successfully implemented, will only serve to anchor the region as a distressed area. Development officials and politicians need to construct a grand alliance with the business sector to promote high-tech manufacturing in the region that pays higher wages than what have normally prevailed in the area. One way this can be accomplished is by revitalizing the union movement in the private sector and respecting wage and benefit levels among public sector unions, rather than looking upon them with envy and as something to be cut. It has long been a complaint among the working classes that the bourgeoisie resists attracting higher-paying industries because such industries would require them to pay more for their own labor, thereby cutting into their profits.

In tracing out the economic transformation of the former anthracite coal regions, a few notable facts stand out. First, these regions are unusual given the fact that they have undergone two waves of deindustrialization. The first was the loss of coal, a result of capital seeking cheaper sources of energy at lower fixed and lower labor costs, an invariable by-product of the process of uneven development in capitalist economies. The second wave of deindustrialization was an emanation of the general processes of deindustrialization that characterized much of industrial America (see Bluestone and Harrison 1982). The coalfields, like other distressed areas in the United States, have tried to recycle their local economies into service economies (see Hyung 2000 for a discussion of Pittsburgh's and Philadelphia's attempts to make this transition). The northern fields, given their greater degree of urbanization (the largest city in the southern/central fields is Pottsville with a population of 14,129 and their initial greater economic diversification have been more successful than the southern fields, although compared to Pennsylvania as a whole, both former anthracite regions are underdeveloped. In part, redevelopment has been impeded by intra- and inter-regional competition, which has been exacerbated by governmental fragmentation and fragmentation of the redevelopment efforts among a number of public and private groups.

Second, mining left behind a large surplus workforce that potential employers saw as ill-suited to modern industrial employment. Miners were seen as lacking the skills necessary for factory work and the labor discipline it required. They were also viewed as an especially unruly lot, given their history of strikes, both authorized and unauthorized. Some of these workers bore the physical costs of employment in the anthracite industry, especially black lung disease, a debilitating and fatal lung disease brought on by years of breathing coal dust. A good many of these miners and, later, unemployed factory workers were older workers whose chances of gaining new jobs were slight. Many of these older workers dropped out of the labor force.

Third, mining helped to produce strong communities and ethnic groups that were committed to the value of a culture of trade unionism and could be mobilized on behalf of labor's causes. This culture showed up in the *Times-Leader/Citizens' Voice* newspaper strike, in which the newspaper unions were able to develop a successful union-owned paper in the face of competition from their former employer, Capital Cities, a large corporate media entity (Keil 1988).

Fourth, as we pointed out earlier, the collapse of mining led to a reworking of traditional gender roles. Large numbers of women were able to emancipate themselves from the old patriarchal family system as they became prime breadwinners and/or took on responsibility of household management.

Fifth, mining left behind a scarred landscape. Culm banks were left behind as markers of what used to be, and underground mine fires had to be contended with, leading to the destruction of at least one community, Centralia, and scarring others; strip mines ripped apart mountainsides; streams and rivers were polluted with mine waste; mine cave-ins became fairly regular occurrences in the former mining zone; and abandoned coal breakers dotted the landscape. The coalfields, in sum, had all of the markings of an industrial wasteland—a place that was outmoded as a production site as a result of capital disinvestments and capital's pursuit of profits in other energy sectors and geographic locales. This scarred landscape is a direct expression of both the years of anthracite dominance and its decline.

Sixth, it is likely that the economic travails produced a number of personal difficulties—stress, anxiety, alcohol and drug abuse, threats to one's identity, and the like. Not surprisingly, little attention was paid to the emotional and human costs of long-term unemployment to the unemployed and their family members when anthracite collapsed.

The overall decline of the region has been repeated in areas other cities northeastern Pennsylvania—Detroit, Youngstown, Flint, Dearborn, and the list goes on. Likewise, other regions with a single industry, be it lumber, textiles, tobacco farming, fishing, iron ore mining, copper mining, and so on, have been devastated in much the same way as northeastern Pennsylvania. All have experienced the effects, principal among which is economic crisis, of capital's inherent tendency to produce uneven development on global, national, regional, and local scales (Smith 2008).

Out of this residue of effects, the coalfields have struggled to find a new economic niche. They thought they had accomplished that with the post-anthracite move to manufacturing, only to find that this strategy would not last. Now, the landscape is speckled with abandoned factories that stand as monuments to a future that never was meant to last. In place of manufacturing, there is a growing service sector that offers lower wages, less career advancement, fewer union protections, and fewer opportunities for economic advancement. Once again the region and its cities have experienced the drastic consequences of uneven development, with its industrial base rendered outmoded by capital flight on a national and global basis. What northeastern Pennsylvania produced could be manufactured at far lower costs and with less regulation elsewhere, and what ensued was another economic crisis for the region.

Northeastern Pennsylvania does not stand alone in experiencing the crisis of deindustrialization. It is one among many regions wounded by processes inherent in capitalism that lead to uneven development across and within regions as capitalists flee "outmoded" production sites in search of new opportunities for higher rates of profit. Such capital flight creates redundant populations that must migrate from their regions of origin or accept a trade down in jobs as their region is devalued in the face of the growth of others.

Bibliography

Adams, Sean Patrick. 1998. "Different Charters, Different Paths: Corporation and Coal in Antebellum Pennsylvania and Virginia." *Business and Economic History* 27 (Fall): 78–90.

American Fact Finder. 2008. *American Fact Finder 3-Year Estimates.* Washington, DC: U.S. Bureau of the Census.

———. 2010. *American Fact Finder 3-Year Estimates.* Washington, DC: U.S. Bureau of the Census.

———. 2012. *American Fact Finder 3-Year Estimates.* Washington, DC: U.S. Bureau of the Census.

Aronowitz, Stanley. 1983. *Working Class Hero: A New Strategy for Labor.* New York: Pilgrim Press.

Aurand, Harold W. 1971. *From the Molly Maguires to the United Mineworkers: The Social Ecology of an Industrial Union.* Philadelphia, PA: Temple University Press.

Bell, Herbert C. 1891. *History of Northumberland County, Pennsylvania.* Chicago: Brown, Runk and Co.

Bimba, Anthony. 1932. *Molly Maguires.* New York: International Publishers.

Blatz, Perry K. 1994. *Democratic Miners: Work and Labor Relations in the Anthracite Coal Industry, 1875–1925.* Albany: State University of New York Press.

Bluestone, Barry, and Bennett Harrison. 1982. *The Deindustrialization of America.* New York: Basic Books.

Bowers, Douglas E. 1983. "From Logrolling to Corruption: The Development of Lobbying in Pennsylvania, 1815–1861." *Journal of the Early Republic* 3 (Winter): 439–74.

Brecher, Jeremy. 1972. *On Strike!* Boston: South End Press.

Bureau of Labor Statistics. May 2010. *Metropolitan and Nonmetropolitan Area Occupational Employment and Wage Estimates. Scranton and Wilkes-Barre, PA.* Washington, DC: U.S. Department of Labor.

Campbell, E. G. 1938. *The Reorganization of the American Railroad System.* New York: Columbia University Press.

Campion, Joan. 1997. *Smokestacks and Black Diamonds: A History of Carbon County, Pennsylvania.* Easton, PA: Canal History and Technology Press.

Chandler, Alfred D., Jr. 1972. "Anthracite Coal and the Beginnings of the Industrial Revolution in the United States." *The Business History Review* 46 (Summer): 141–81.

City-data. 2013a. Crime Rate in Carbondale, Pennsylvania. http//www.city-data.com/crime/crime-Carbondale-Pennsylvania-html. Accessed September 6, 2013.

———. 2013b. Crime Rate in Hazleton, Pennsylvania. http//www.city-data.cocm/crime/crime-Hazleton-Pennsylvania.html. Accessed September 6, 2013.

———. 2013c. Crime Rate in Nanticoke, Pennsylvania. http//www.city-data.com/crime/crime-Nanticoke-Pennsylvania.html. Accessed September 6, 2013.

———. 2013d. Crime Rate in Pottsville, Pennsylvania. http//www.city-data.com/crime/crime-Pottsville- Pennsylvania.html. Accessed September 6, 2013.

———. 2013e. Crime Rate in Scranton, Pennsylvania. http//www.city-data.com/crime/crime-Scranton- Pennsylvania.html. Accessed September 6, 2013.

———. 2013f. Crime Rate in Shamokin, Pennsylvania. http//www.city-data.com/crime/crime-Shamokin-Pennsylvania.html. Accessed September 7, 2013.

———. 2013g. Crime Rate in Sunbury, Pennsylvania. http//www.city-data.com/crime/crime-Sunbury_Pennsylvania.html. Accessed September 7, 2013.

———. 2013h. Crime Rate in Wilkes-Barre, Pennsylvania. http//www.city-data.com/crime/crime-Wilkes-Barre-Oennsylvania.html. Accessed September 7, 2013.

Coleman, J. Walter. 1936. *The Molly Maguire Riots: Industrial Conflict in the Pennsylvania Coal Region.* New York: Arno Press.

Corbin, David Allen. 1981. *Life, Work, and Rebellion in the Coalfields: The Southern West Virginia Miners, 1880–1922.* Urbana: University of Illinois Press.

County Profiles. 2009. Harrisburg, PA: Center for Rural Pennsylvania.

Cox, Oliver Cromwell. 1948. *Caste, Class, and Race.* New York: Doubleday.

Crown, H. T., and Mark T. Major. 2003. *A Guide to the Molly Maguires.* Frackville, PA: Black Mountain Publishing.

Cumbler, John T. 1979. *Working Class Community in Industrial America: Work, Struggle, and Leisure in Two Industrial Cities, 1880–1930.* Westport, Conn.: Greenwood Press.

Cummings, Scott. 1980. "Collectivism: The Unique Legacy of Immigrant Economic Development." In *Self-Help in Urban America: Patterns of Minority Economic Development,* edited by Scott Cummings. Port Washington, NY: Kennikat, 5–29.

Davies, Edward J., II. 1985. *The Anthracite Aristocracy: Leadership and Social Change in the Hard Coal Regions of Northeastern Pennsylvania, 1800–1930.* DeKalb: Northern Illinois University Press.

Dublin, Thomas, and Walter Licht. 2008. "Gender and Economic Decline: 1920–1970." In *Backcountry Crucibles,* edited by Jean R. Soderlund and Catherine S. Parznski. Bethlehem, PA: Lehigh University Press, 329–43.

Ernst, Morris L. 1937. *Anthracite Coal Commission Report.* Harrisburg: Commonwealth of Pennsylvania.

Folsom, Burton W., Jr. 1980. "Like Fathers, Unlike Sons: The Fall of the Business Elite in Scranton, Pennsylvania, 1880–1920." *Pennsylvania History* 47 (October): 291–309.

———. 1981. *Urban Capitalists: Entrepreneurs and City Growth in Pennsylvania's Lackawanna and Lehigh Regions, 1800–1920.* Baltimore: Johns Hopkins University Press.

Goin, Peter, and Elizabeth Raymond. 2001. "Living in Anthracite: Mining Landscape and Sense of Place in Wyoming Valley, Pennsylvania." *The Public Historian* 23 (Spring): 29–45.

Gordon, David M., Richard Edwards, and Michael Reich. 1982. *Segmented Work, Divided Workers: The Historical Transformation of Labor in the United States.* New York: Cambridge University Press.

Greene, Victor. 1968. *Slavic Community on Strike.* South Bend, IN: University of Notre Dame Press.

Hanlon, Edward F., and Paul J. Zbiek. 2003. *The Wyoming Valley: An American Portrait.* Sun Valley, CA: American Historical Press.

Healey, R. G. 2007. *The Pennsylvania Anthracite Coal Industry 1860–1902: Economic Cycles, Business Decision-Making, and Regional Dynamics.* Scranton, PA: University of Scranton Press.

Hirsch, Barry, and David Mcpherson. 2010. "Union Membership and Coverage Database from the CPS," Unionstats.com.

Hobsbawm, Eric. 1959. *Primitive Rebels.* New York: Norton.

Howard, Walter T. 2001. "The National Miners Union: Communists and Miners in the Pennsylvania Anthracite, 1918c–1931." *Pennsylvania Magazine of History and Biography* 125, no. 1–2: 91–124.

Hyung Je Jo. 2000. *Regional Restructuring and Urban Regimes: A Comparison of the Pittsburgh and Detroit Metropolitan Areas.* Ann Arbor: University of Michigan Transportation Research Institute, Office of the Study of Automotive Transportation.

John, J. J. 1891. "Development of the Shamokin Coalfield." In *History of Northumberland County, Pennsylvania,* edited by Herbert C. Bell. Chicago: Brown, Runk and Co. 358–91.

Kanarek, Harold K. 1974. "Disaster for Hard Coal: The Anthracite Strike of 1925–1926." *Labor History* 15 (Winter): 44–62.

Keil, Thomas J. 1982. "Capital Organization and Ethnic Exploitation: Consequences for Miner Solidarity and Protest." *Journal of Political and Military Sociology* 10 (Fall): 237–55.

———. 1984. "Mobilizing Adolescent Support for an American Newspaper Strike: Results from a Case Study." *Organization Studies* 5/4: 327–43.

———. 1988. *On Strike: Capital Cities and the Wilkes-Barre Newspaper Unions.* Tuscaloosa: University of Alabama Press.

Lauck, W. J., and Edgar Sydenstricker. 1917. *Conditions of Labor in American Industries.* New York: Arno.

Lewis, Arthur H. 1964. *Lament for the Mollie Maguires.* New York: Harcourt Brace and Jovanovich.

Luther, P. D. 1881. "Development of the Coal Production and Trade in Schuylkill County." In *History of Schuylkill County, PA.*, edited by W.W. Munsell. New York: Press of George Macnamara, 41–72.

Macke, Don, Ahmet Binerer, Taina Radenslaben, Dick Gardner, Tim Murphy, and Eric Thompson. 2008. "Wealth Transfer in Pennsylvania." *Rural Perspectives* (January–February 2011): 5–24. Harrisburg.

Marsh, Ben. 1987. "Continuity and Decline in the Anthracite Towns of Pennsylvania." *Annals of the Association of American Geographers* 77 (September): 337–52.

McMillion, George W. 2008. *Pennsylvania's Jobs Market 2001–2008: Trading Down in a Debt-Driven Weak Economy.* Washington, DC: American Manufacturing Trade Action Coalition.

Miller, Donald L., and Richard E. Sharpless. 1985. *The Kingdom of Coal: Work, Enterprise, and Ethnic Communities in the Mine Fields.* Philadelphia: University of Pennsylvania Press.

Murphy, Raymond E., and Marion Murphy. 1938. "Anthracite Region of Pennsylvania." *Economic Geography* 4 (October): 338–48.

Nelson, Steve, James R. Barrett, and Rob Ruck. 1981. *Steve Nelson, American Radical.* Pittsburgh: University of Pittsburgh Press.

Novak, Michael. 1978. *The Guns of Lattimer.* New York: Basic Books.

O'Bannon, Patrick, Martin B. Abbott, Susan C. Nabors, and James T. Parkinson. 1997. *Anthracite Coal in Pennsylvania: An Industry and a Region.* Harrisburg: Bureau of Historical Preservation, Pennsylvania Historical and Museum Commission.

Oppenheimer, Martin. 1974. "The Sub-Proletariat: Dark Skins and Dirty Work." *The Insurgent Sociologist* 4 (Winter): 6–20.

Pennsylvania Bureau of Statistics and Planning. 1974. *Pennsylvania Abstract.* Harrisburg: Pennsylvania Bureau of Statistics and Planning.

Pennsylvania Keystone Opportunity Zone/Keystone Opportunity Expansion Zone. 2003. KOZ/KOEZ Four-Year Report. Harrisburg: Commonwealth of Pennsylvania.

Pennsylvania Manual. 1979. *The Pennsylvania Manual* vol. 104. Harrisburg: Department of General Services, Commonwealth of Pennsylvania.

Peters, Alan H., and Peter S. Fisher. 2002. *State Enterprise Zones: Have they Worked?* Kalamazoo, MI: W.E. Upjohn Institute for Employment Research.

Phelan, Craig. 1994. *Divided Loyalties: The Public and Private Life of Labor Leader John Mitchell.* Albany: State University of New York Press.

Record Year Book and Almanac. 1911–1945. Wilkes-Barre, PA: Wilkes-Barre Record Company.

Reich, Michael. 1977. "The Economics of Racism." In *Problems in Political Economy,* 2nd ed., edited by David M. Gordon. Lexington, MA: D.C. Heath, 183–87.

Roberts, Peter. 1901. *Anthracite Coal Industry.* New York: Macmillan.

Rochester, Anna. 1931. *Labor and Coal.* New York: International Publishers.

Smith, Neil. 2008. *Uneven Development: Nature, Capital, and the Production of Space,* 3rd ed. Athens: University of Georgia Press.

Stepenoff, Bonnie. 1999. *Their Fathers' Daughters: Silk Mill Workers in Northeastern Pennsylvania, 1880–1960.* Selinsgrove, PA: Susquehanna University Press.

Stolarik, Mark. 1980. "A Place for Everyone: Slovak Fraternal-Benefit Societies." In *Self-Help In Urban America: Patterns of Minority Economic Development,* edited by Scott Cummings. Port Washington, NY: Kennikat, 130–44.

Thompson, Fred W., and Patrick Murfin. 1976. *The IWW: Its First Seventy Years, 1905–1975.* Chicago: Industrial Workers of the World.

U.S. Anthracite Coal Strike Commission. 1903. *Report to the President on the Anthracite Coal Strike of May–October, 1902.* Washington, DC: U.S. Anthracite Coal Strike Commission.

U.S. Bureau of the Census. 1990. *U.S. Census of Population and Housing, 1990.* Washington, DC: U.S. Bureau of the Census.

———. 1997. *Selected Economic Characteristics, 1997.* Washington, DC: U.S. Bureau of the Census.

———. 2000. University of Virginia Library Geostat Collections, http://www.fisher.lib.viginia.edu/collections/stats/histCensus/index.html.

———. 2007. *Selected Economic Characteristics, 2007.* Washington, DC: U.S. Bureau of the Census.

———. 2010. *U.S. Census of Population and Housing, 2010.* Washington, DC: U.S. Bureau of the Census.

———. 2012. *American Community Survey.* Washington, DC: U.S. Bureau of the Census.

United States Geological Survey. 1997. *The Use of Historical Production Data to Predict Future Coal Production Rates,* http//www.pubs.USGS.give/circular/c1147/historical.html.

United States v. Reading Company et al. 1912. U.S. 26, 33 Sup. Ct. 90, 57 L.Ed. 243 (1912).

———. 1920.253 U.S. 26, Sup. Ct., 64, L. Ed. 760.

University of Wisconsin Population Health Institute. 2010. *County Health Rankings: Mobilizing Action toward Community Health.* Madison: University of Wisconsin.

Wardell, Mark, and Robert L. Johnston. 1987. "Class Struggle and Industrial Transformation: The U.S. Anthracite Industry, 1820–1902." *Theory and Society* 16 (November): 781–808.

Wolensky, Kenneth C., N. H. Wolensky, and R. P. Wolensky. 2002. *Fighting for the Union Label: The Women's Garment Industry and the ILGWU in Pennsylvania.* University Park: Pennsylvania State University Press.

Wolensky, Robert P. 1993. *Better Than Ever! The Flood Recovery Task Force and the 1972 Agnes Disaster. Stevens Point:* Stevens Point: University of Wisconsin—Stevens Point Foundation Press.

Wolensky, Robert P., K. C. Wolensky, and N. H. Wolensky. 1999. *The Knox Mine Disaster.* Commonwealth of Pennsylvania. Harrisburg: Pennsylvania Historical and Museum Commission.

Yancey, William L., Eugene P. Erikson, and Richard L. Juliani. 1976. "Emergent Ethnicity: A Review and Reformulation." *American Sociological Review* 41 (June): 391–403.

Yearly, Clifton K., Jr. 1961. *Enterprise and Anthracite: Economics and Democracy in Schuylkill County, 1820–1875.* Baltimore: Johns Hopkins University Press.

Index

About the Authors

Jacqueline M. Keil is currently a political science faculty member at Kean University in Union, New Jersey. She received her MA and PhD from The Ohio State University. Her doctoral dissertation focused on international economic, political, and organized labor issues. Dr. Keil's current research and teaching interests are the Tea Party movement, how the Occupy Wall Street movement addresses economic inequality in the United States, human rights, the role of social media in politics, and the politics of popular culture.

Thomas J. Keil is professor emeritus of sociology at Arizona State University, where he also served as associate dean of the New College of Interdisciplinary Arts and Sciences. Dr. Keil received his undergraduate degree from King's College in Wilkes-Barre, Pennsylvania, and his MA and PhD from Temple University. Before coming to Arizona State he previously taught at Gettysburg College and the University of Louisville. From 1991 to 1992, he was a Fulbright Scholar at the University of Bucharest and conducted research at the Center for Regional and Urban Sociology in Bucharest and at the Romanian Academy's Center for the Study of the Quality of Life, also in Bucharest. His research has focused largely on capital punishment and political economy.